THE ENJOYMENT OF CHEMISTRY

The Enjoyment of
CHEMISTRY

Louis Vaczek

Foreword by René J. Dubos

The Viking Press · New York

FIRST PUBLISHED IN 1964 BY THE VIKING PRESS, INC.
625 MADISON AVENUE, NEW YORK, N.Y. 10022

PUBLISHED SIMULTANEOUSLY IN CANADA BY
THE MACMILLAN COMPANY OF CANADA LIMITED
SECOND PRINTING JUNE 1965

LIBRARY OF CONGRESS CATALOG CARD NUMBER: 64-11193

ENDPAPERS PHOTOGRAPH
BY GENERAL ELECTRIC RESEARCH LABORATORY

PRINTED IN THE U.S.A. BY THE VAIL-BALLOU PRESS, INC.
MBG

To my sons, Nicholas and Adam

Foreword by René J. Dubos

Almost every day newspapers publish accounts of the prodigious chemical reactions by which laboratory scientists convert the crude components of matter into the sophisticated and glamorous products of modern technology. The accounts are generally correct, because science writers now have a broad theoretical training and are careful in checking the accuracy of their reports. But the newspaper accounts do not tell the whole story, because the most exciting reactions occur not in test tubes or factories, but in the mind of the scientist while he holds the test tube or designs the factory.

Mr. Vaczek has successfully carried out a very bold task. He has translated into clear and simple words the thought processes which chemists use in their attempts to describe matter—including the matter of which living things are made—and to transform matter for the benefit or entertainment of mankind. Moreover he has traced the chain of reactions which has linked into a common structure the minds of scientists all over the world throughout ages—from the time when Democritus imagined that matter was made up of hard particles which he called atoms, through the fateful days when Einstein showed that matter could be converted into energy, to the mysterious and awe-inspiring activities of the theoreticians who now try to discover the ultimate particles of the natural world.

The facts and theories of which Mr. Vaczek writes are complex, indeed beyond human experience. But the ideas themselves are paradoxically simple. The reason is that chemists are, after all, just plain ordinary men. Their thought processes are fundamentally the same as those of other men. And what Mr. Vaczek has done is to take his readers by the hand, to lead them through the channels of thought which the human scientific collectivity has followed for several thousand years. His book will not transform laymen into chemists, but it will make it possible for all intelligent persons to share in the intellectual enjoyment of those who probe into the structure of matter and the machinery of life.

Contents

PART III

THE CHEMICAL NATURE OF ATOMS

THE ENJOYMENT OF CHEMISTRY

PART I

Atoms Unlimited

•-•

1

"Things Are Seldom What They Seem"

Man's sorting mind and language

Do man's sensory impressions provide him with a true idea of the real world, or do his impressions create a dream of some world that exists only in his mind? From earliest history philosophers have debated this question. The scientific view brushes aside the question and flatly assumes that a real world exists and that man is real in it and that man's impressions of that world are true and therefore yield him a real knowledge of nature.

The most primitive mind recognizes similarities and contrasts and patterns of events in the external world that are essentially the same as the patterns recognized by all minds, and thus the idea of a universal order dominates every culture. Furthermore, every generation teaches its children that some kind of cause-

and-effect sequence underlies the patterns of everyday experience: things happen for a reason, and man is confident that therefore he can plan for the future. Even a baby hears his mother's croon as part of a larger pattern of similarities and contrasts that include his hunger pangs, crying, a door opening, and so on.

The eyes of any man can see that one leaf on an oak tree resembles every other leaf on that tree and resembles the leaves of all oaks anywhere and, compared to any other kind of thing, an oak leaf anywhere resembles the leaves of all trees that have ever grown. For this undefinable leaf shape, leaf texture, leaf function—compared to all roots, or all fish, or all clouds—for this abstract idea that only the mind sees fluttering at all the imagined twig-ends of the world, the mind invents a single, symbolic word: leaf. For all trees of any kind whatsoever: tree.

The youngest, the most primitive, the simplest mind observes the uncountable number of things in the world and their infinite variety, and, instead of trying to identify each one with a different name, the mind reduces the infinite variety and numbers to a few groups: leaf, sand, water, cloud, fox, man.

But the members of a group are recognizable because each one has a color, hardness, heaviness, wetness, and so on, similar to the color, hardness, wetness, and so on, of the other members of that group, but not similar to the color, hardness, and so on, of the members of other groups. Thus we have created a category of abstractions at the same time that we created the category of objects: their properties whereby we recognized them. To see the color red in an apple, a friend's hair, a flame, and to say that those different objects are red, is an automatic function of the mind, it seems. Without this ability we would be unable to distinguish one object from another. It is not the actual objects that we experience, but the variety of impressions that their properties make on us.

Both these concepts—the concept of objects and the concept of properties—have been frustrating philosophers from the earliest times. Here we wish only to point out that language is a symbolic record of the mind's ability to create categories of impressions and that with these categories the complexity of the universe is enormously simplified.

If we now try to describe our universe using only the words for objects and their properties, we will create a totally motionless and soundless, and, strictly speaking, a totally unlighted, landscape. Nothing at all is happening to those objects, to those nouns and descriptive adjectives.

All language contains still another category of abstractions symbolized by verbs that relate to the actual behavior of objects. Since this is not a treatise on linguistics, we use the word "symbol" in its most general sense, without intending to obscure the different and precise meanings of connoting, denoting, expressing, representing, signaling and so on. Let us say that predicates are concerned with action, change.

We find that there are pressures, temperatures, the pull of gravity, the push and pull of magnetism or of electricity, the insistence of light, and forces of many other kinds that act on our objects to change their positions and their shapes, and even literally to transform them from one kind of substance into another. It is the countless kinds of unceasing change in our landscape, and in ourselves, that create the primary sense of existence: birth, growth, decay, the seasons, the tides, the circling sky, fire, thought itself, death.

All things change but not uniquely once and forever, and we can abstract categories of change. The word "death" does not stand just for a familiar, endlessly repeated event; it symbolizes a concept of a force infinitely superior to any individual death and it relates far more profoundly to the concept of another force symbolized by the word "life" than it does to any individual life.

When we consider these actions and transformations carefully they seem to be imposed upon our world. We are acted upon, we are influenced, tempted, pushed, beguiled, horrified; we respond far more often to our environment than at first glance we think we do. The more we probe an event, the more clearly do we see that primal forces initiated the action, primal forces that did not originate in the objects affected. In the vegetable and inanimate worlds almost no initiative can be found, and yet everything is changing there too. The origin of change seems to be independent of the palpable world.

All the early records of man reveal that he always discovered the same universal changes and distinguished them from the natural properties of the material world, and that then he invented a supernatural world in which those forces originated. The real world of every day seems to the primitive mind to be in perpetual thrall to metaphysical powers which control all change and movement. These invisible powers seem to manifest their presence in thunder and in flowering trees, in love and hate, birth and death, sickness, the wind.

Thus, each culture has invented its own spirits and gods who manipulate the world, and has invented its own method of communicating man's wishes concerning his world to those immaterial powers. All men everywhere yearn to rule their environment, and to rule they must somehow plan for the future. Every culture has trained its own style of witch doctors, priests, magicians, scientists, to impose man's will upon the forces of nature by prayer, by magic, or by microscope and test tube.

The first organizations of information

The people in the West whose civilization was the first to include a devoted, cooperative effort to collect information about nature with the aim of uncovering nature's orderly processes were the Babylonians, who were also the first to imagine and record a cosmology, an orderly or systematized creation. Standing on religious ground, about four thousand years ago they began to collect scrupulously details about the movement of the stars and planets and eventually perceived what they considered to be supreme cyclical patterns of the universe at large. By the third century B.C. the Babylonians believed that these cosmic cycles affected everything here on earth too, including the life of each human being and animal. The stars, in their view, were the source of all change, and because the stars changed their positions in orderly fashion changes here on earth could be predicted. The longing to control environment created Babylonian astrology, which to this day is considered an important body of prophetically useful knowledge by many people, whatever we may think of astrology scientifically.

But it was the priests of ancient Egypt who were the first to develop a comprehensive approach to the transformation of one substance into another, and since theirs was also a wholly religious approach the knowledge they collected is expressed largely in religious terms. Not only the temple mysteries but all the crafts of the Nile were rooted in a metaphysical view of matter that was, at the same time, concerned with practical achievements in the real world. The dyeing of cloth, the manufacture of glass, the smelting of metals from ores, the tanning of leather, the production of medicines and poisons, the practices of farming, the recipes of cooking, and the mummifying of a corpse are all chemical techniques for transforming one kind of substance into another. In Egypt high standards were demanded because of the religious use of objects, and therefore formal procedures were worked out in all the crafts and taught century after century to apprentices. Though religious beliefs changed with the movement of people, concepts concerning the actual nature of matter did not change. The Egyptians did not speculate about the clay of reality: that clay seemed to be in the hands of the gods.

In the sixth century B.C. Greece, where philosophy and not religion dominated intellectual activity, produced the first men to become fascinated by the structure of matter itself. Unlike the Egyptians, the best Greek minds were not awed by the marvels of handicrafts that any slave could be taught. They drew off and observed nature detachedly, without desire for technically useful knowledge but with a longing for ultimate truths.

Among the first ideas Greek philosophers discussed was the intuitive one that, since the actual shape of things could be altered by hacking and chopping, the shape and appearance of objects was a transient manifestation of something more fundamental than either shape or appearance. Perhaps, they speculated, there existed an underlying prime stuff from which all things were fashioned. One philosopher concluded that air itself was the universal element that, in various degrees of compression, constituted the varieties of substances. No, said another, the basic stuff was water. Fire, said a third. Then Empedocles proposed not one primeval fundamental substance or element but four: air, water, earth, and fire. All things, he said, were a mixture of these four elements in various proportions, and everything, including cabbages, sealing wax, and the different parts of kings, could be broken down into air, water, earth, and fire.

Of course these ideas were not all discussed at once for the first time one fine afternoon in a garden full of philosophers who had met just for that purpose. Generations came and went while the theories were developed by different teachers and later taught in different schools.

About 420 B.C. the Greek philosopher Democritus took a completely fresh line of approach to the question: what is matter? Democritus and his teacher Leucippos fastened their imaginations on the problem of whether or not matter was continuous. Can one imagine cutting any object into smaller and smaller pieces and never arriving at a final, ultimate particle that cannot be split any more? Or is it possible to conceive of an ultimate particle which, under a final blow, does not split in two but simply vanishes into nothing? Is it possible, in other words, to destroy all the substance of an apple, simply by chopping away at it long enough? Democritus, in line with his teacher's idea, concluded that matter had to be discontinuous, that is, made up of finite, invisibly small but indestructible "atoms"—his word was *atomon*, meaning indivisible —and that the atoms of each kind of substance had their own particular shape and texture.

Among the questions his hypothesis raised was: how do atoms congregate and cling together to produce the large objects we know? They are in constant motion, said Democritus, and they love or hate, attract or repel one another in specific ways; their coming together creates the bulk matter that is familiar to us.

He was denounced for religious reasons. In a universe such as he proposed all would be chance, a random series of clingings and repulsions, for we cannot imagine that each tiny atom is imbued with conscious will and cognizance of its role. A universe of brute atoms winging about in random manner would have no moral purpose.

Democritus had few followers in Greece, and four centuries later in Rome another philosopher, Lucretius, argued fruitlessly for the same theory. During the Middle Ages now and then a voice was raised defending atomism. But soon after Democritus the giant intellect of Aristotle had effectually demolished the theory, and with it all the questions that it asked and could not answer.

The unique rule of Aristotle

Aristotle enshrined the theory of the Four Elements as the truth about matter, together with two properties, or qualities, for each. Water was cold and wet, earth was cold and dry, air was wet and hot, fire was hot and dry. The various amounts in which the elements were mixed produced the material substance of an object, while the various amounts in which their attributes, the four qualities of wet, cold, dry, and hot, were mixed produced the properties of the substance. He added a fifth element, a quintessence—aether, first suggested by the Pythagoreans as the substance out of which the celestial bodies are made.

This Aristotelian structure of abstract qualities and primal elements ruled the West's ideas of matter for the next 2000 years. It was an integral part of an orderly model of creation in which man and his mind were given a noble role, above the beasts but below the angels, and in which every part emphasized the common sense of ordinary experience. Since this mode of thought put man at the center of an ordered universe and mixed philosophy and science, no part of the model could be changed without changing all of it—really, without destroying it.

But the long rule of Aristotle is due partly to the fact that his is the most comprehensive "system" of thought in history, East or West. Not even modern science is yet as comprehensive. And it is also due partly to the fact that, if we ignore experimentation, the mind can even today accept what Aristotle stated about the world, because he relied on the senses and on common sense. And partly, perhaps largely, Aristotle's rule lasted so long because the Christian Church made the Aristotelian cosmogony its own and all education in Europe was in the hands of the Church, which to this day maintains the conviction that man has a key position in the real universe.

It was not until the seventeenth century that a system of thought, a methodology, an attitude, a point of view, a logistics of reasoning, as it were, developed into the first principles of our scientific approach to reality, an approach that brushed aside moral speculation inside the laboratory. Therefore the power of the Four Elements must be seen in their relation not only to ancient philosophy but to Christianity as well.

Alchemy, or the black art of Egypt

When Athens decayed, the Aristotelian cosmology migrated to Alexandria, founded in Egypt three centuries before Christ, and there the world-famous Four Elements—philosophical, abstract, universal, absolute—met the world-famous practical technology of the Egyptian crafts and priest-rule. The purely theoretical met the purely empirical and fused into the origin of alchemy, a proto-science whose name was coined for it later on by the Arabs. Since everyone was well versed in Babylonian astrology, its principles were poured right into the first mix, and eventually alchemical practices absorbed Platonic ideas, numerology, Christian and Hebrew mysticism, local and international abracadabra, witchcraft, necromancy, folk lore, pagan mythology, an amazing number of close observations of nature, and all the undigested and undigestible knowledge and guesses of the ages, concerning chemical reactions.

It is impossible to define alchemy because it is difficult to narrate its history, because the records it left are sparse and cryptic, and because at one time or another every famous medieval philosopher, scholar, and mystic had something to do with the body of information that was considered to be alchemical. A brief outline of the general reasoning that motivated alchemical activity may be imagined, though such an outline would not have occurred to any alchemist, and in fact it would have been rejected outright by all of them. Hindsight is one of the most useful tools the mind has, to put it mildly.

The universe, begins the reasoning, was created out of a mixture of four (or five, or, according to some alchemists, seven or even nine) primary elements. The enormous variety of substances is due to the possibility of mixing in endless proportions the attributes of the elements with certain other varieties of essences, aethers, forces, spirits, mysteries. Most substances as they now appear on earth are mixed with impurities but they can be refined, or purified, into their original pristine condition. In other words, a mixture can be broken down into its elements.

The hierarchy of power and beauty that places man below the angels, and all other creatures below man in an order of descending values, is continued in the inanimate world. Some substances by their intrinsic nature are finer than others. The finest of all and the most noble, because it remains unchanged under all conditions, is gold. Just as man aspires to higher estate and achieves it on a spiritual level, so do the coarse substances yearn to be noble. Lead is willing, in a manner of speaking, to be transformed into gold. All things can be transformed in due processes of nature. The meat of animals and the substance of plants, when eaten by man, become man.

Since there is a prime substance that underlies everything, changes in matter are merely changes in the ratio of qualities, or properties, of the prime stuff. Lead is dense and malleable but dull, dark, and poisonous, while gold is just as dense and malleable but lustrous like the sun. If that gray could be altered to yellow, then the prime stuff in lead would have acquired the properties of gold.

After all, visible transformations of matter are going on constantly. One can call such transformations "transmutation of the elements."

The agencies in these transmutations are mysterious but can be controlled by those who know how to use the right methods. How wonderful it would be if a reagent existed that could transmute everything base into its ideal, pure state, for instance, base metals into gold—a Philosopher's Touchstone! And what about something that did the same for living things—an Elixir of Life which could rejuvenate bodies, washing away the impurities of aged flesh and transforming it into the glorious shapes and colors of youth? And a Universal Solvent would certainly be helpful in the process of purification. Since these reagents would not only be marvelous if they existed but also bring their possessor untold wealth, it was worth while searching the earth, the air, even heaven and hell for them.

Such was, and always will be, the trap of wishful thinking. It is part of man's nature, his inalienable right, as it were, to think that the world can be improved, and all we can do is guard and painstakingly sort out sparse, genuine facts from the continuous flow of hopeful ideas about facts.

Anthropologists point out the profound need in every culture for self-purification through ritual. Christian beliefs concerning sin and redemption hovered over alchemical thought concerning chemical change and the transmutation of the elements. Both chimed with Aristotle's cosmology. The soul, the essence of anything, could be lost to the Powers of Darkness or redeemed by the Powers of Light. The concept of purification was therefore applicable on both the material and spiritual planes.

To sum up, for alchemists the surface appearance of matter was one thing and its essential substance was another thing. This is also the view of most philosophies, and it is certainly the view of modern science. In the past, endless experiments with the surface appearance of things led to remarkable reasoning from one erroneous theory to another. The most brilliant men of the Middle Ages and of the Renaissance, and even some in the nineteenth century, accepted what seem now to be wild fantasies concerning the behavior of matter, for they could not resist the pleasure of imagining a perfect universe shimmering just beyond this graveled reality. And there was something wonderful in believing that four or five or seven or nine universal elements, perfect abstractions, constituted all matter.

Why was the rule of alchemy never challenged when intelligent men recognized that it was a cloak for charlatans who seduced even kings with promises to manufacture gold? For the simple reason that alchemists were the chemists and the pharmacists, who brewed medicines and poisons and worked with acids and metals as a matter of course. They were adept in all the "natural sciences" of astronomy, mathematics, and physics; they were physicians, teachers, scholars in general. The courts of Europe and rich men employed alchemists for practical purposes, and the most honest was expected to keep his ears open for rumors of new cure-alls. No man really knew, after all, that the Philosopher's Stone did not exist, or that lead could not be transmuted into gold. Furthermore, knowledge itself was desirable in those days, just as it is now. There is a lust for knowledge in all men, against the tedium of brute existence.

The written record of alchemical activity is amazingly slim when we consider its nearly two thousand years of life. It is interesting to speculate on the reasons.

First, the practical technology inherited from Egypt was soaked with hocus-pocus that, even when it was ignored, encouraged the feeling of cultism. A cult guards secrets. The profit motive also made each alchemist guard his own ideas and discoveries jealously. Then, legal and religious persecution, from Roman days on, of magicians and witches drove scholarly researchers underground from time to time, fearful of being caught with incriminating documents of a supernatural cast. Anyone who has passed the windows of a high-school laboratory can imagine how the unventilated cellars of alchemists must have reeked and how even the clothing of the wizard as he passed in the street must have wrinkled the noses of God-fearing citizens and made him proud and silent.

Such substances as bitches' milk and whites of egg and excrement were used in experiments, as though their actual composition were irrelevant compared to their essence, and the idea of what their essence was varied from brain to brain. There was no general agreement on anything, not on how to set up and observe and report an experiment, not on how to measure ingredients. A pinch of salt, a small shaving of copper, a few drops of strong acid, heat until the mixture boils, then add a cupful of ashes— this is more like a private recipe than an orderly procedure in a disciplined sequence of operations that everyone could repeat. Even if alchemists had had international publications and annual conferences to exchange information, they could not have communicated much more than they did in letters to one another for the simple reason that their basic assumptions concerning matter were variations on Aristotle's statements.

In any case, through those centuries not much was committed to the record about anything, let alone alchemy, and whatever was written of alchemical theories was mostly in cryptic, cabalistic, symbolic, mystical language that discouraged all but the hardiest seeker. The whole history of alchemy seems to have been one of solitary lust for fortune at one extreme and solitary lust for knowledge at the other extreme.

Nevertheless the surprisingly meager list of things that we inherited from the alchemists is quite significant. First is the very idea of industrious, long-range, planned experiments in a laboratory set aside from other places and equipped with special apparatus. Some of the apparatus in use today was designed a thousand years ago; for instance, the crucible for fusing metals and the still for separating mixtures of liquids. In their laboratories the alchemists developed methods of manufacturing some of our important reagents, such as sulfuric and nitric acids, and they discovered quite a few of the true elements and the compounds that these form. In metallurgy the knowledge they collected was pure enough to have served us well until today. But most important, perhaps, is their use of symbols and formulas instead of words, and the way in which they included these symbols in mathematical operations as though they were numbers. It is ironical and entertaining to realize that the same magic signs which concealed an alchemist's knowledge today enable chemists to handle far more knowledge than ordinary speech could.

Alchemy was kept alive not by fools, lunatics, and thieves but by the best minds, century after century trapped in the network of impressions that the senses feed to the mind, impressions which, untested by experimentation, led inevitably to dreamlike beliefs concerning matter. It was all part of our old civilization, no better and no worse than its usefulness.

Phlogiston, the burning principle

The end of alchemy came with the exploding of a wonderful theory concerning fire, about which countless theories had been proposed since the first legends—a theory that illustrates the enormous difference between pre-nineteenth-century and post-nineteenth-century approaches to chemical change.

In the eighteenth century nearly all scientists accepted the common-sense conclusion based on simple observation that, when a substance burns, part of it is lost in smoke and flame, while the rest, the remainder, is reduced to ashes. Something fiery escapes from burning matter, ran the reasoning. This something was named *phlogiston*, a Greek word meaning "inflammable." Anything that could burn contained phlogiston, so that a piece of wood consisted, really, of wood ashes and smoke combined with phlogiston.

Adjustments had to be made to this theory to explain the great variations in the burning processes, some of which are not as simple as that of wood, until phlogiston acquired such a complex nature that at times it was supposed to have less than no weight. For instance, when a metal is burned, or calcined, the resulting ash, or calx, actually weighs more than the original metal did.

Obviously, went the explanation, when combined with a metallic ash, phlogiston lightens the calx, which becomes the metal, and when phlogiston escapes the metal in the flames, it turns into the heavier calx again. But such contradictions, far from discrediting the theory, gave it mysterious charm.

Of course by the end of the eighteenth century many chemists had struggled free of traditional beliefs and were experimenting openly and intelligently and exchanging information through scientific societies founded in the previous century. The sheer passion to find out how the world was really made by actually picking it apart had been born soon after Aristotle's heavenly spheres were shattered by Galileo's telescope and Kepler's great discovery of the elliptical orbits in which the planets move around the sun. All Aristotle's ideas became suspect at once and were put to the test one by one with devices and instruments he had never dreamed of to aid the senses. Knowledge of a kind that no one could ever discover by reasoning alone flooded men's minds. Consequently, many thoughtful experiments with fire were made and, though the observations were twisted to fit the phlogiston theory, they were recorded plainly enough as bald facts.

The father of modern chemistry

At last a Frenchman, Antoine Lavoisier, repeated an experiment in burning that had been described to him, gave the correct explanation of what had happened, and thus exploded the theory of phlogiston and everything else that was left of alchemical mysteries. This historic experiment was conducted just before the French Revolution broke out, when Lavoisier, already a famous scientist, was hard at work to establish gravimetric procedure in the laboratory as the fundamental principle for all chemical knowledge. ("Gravimetric" means that everything is weighed or measured.) It is a sobering thought that he was opposed by most of the distinguished chemists of his day.

Lavoisier proved that, instead of phlogiston's escaping when a substance burns, the substance combines with a part of the atmosphere so that the ashes and the smoke, which can be easily trapped in a simple apparatus during a properly designed experiment, weigh more than the original unburned substance did. If all the hot gases rising from a burning candle are caught and weighed, they will weigh more than the original candle did. Air, said Lavoisier, basing his conclusion on weighings he made in a variety of experiments in burning, is a mixture of gases, one of which combines chemically with the burning substance. This gas must be present in the air, or the burning process, also called combustion, will not take place. When this gas has been used up, as, for instance, by burning a candle in a sealed jar of air, the burning stops. Lavoisier named this gas "oxygen." From other experiments he drew the conclusion that respiration was also a process of combustion. A living animal draws oxygen into its lungs, where oxygen combines with organic material, and it breathes out the products of combustion: water and carbon dioxide, the same substances that burning wood produces.

This correct explanation of fire and of life was the cornerstone on which modern chemistry proceeded to build. Lavoisier actually used mercury in his original experiment, and it is entertaining to note that a metal named after the messenger of the gods informed him of the true nature of fire, man's oldest fascination.

In 1794 Lavoisier was guillotined by the Reign of Terror as an enemy of the people, because he had a job with the tax-collecting department.

2

Modern Science Begins with Measurement

History is a point of view

To follow strict historical development would obviously challenge the reader with the same problems that puzzled brilliant scientists—problems that were not solved for centuries, always with great crisscrossing of ideas and slow digestion of data and a confusion of beliefs in this or that assumption, and very often with the elimination of common sense or of some axiom that until then seemed to be a keystone of truth.

No one quite knows how to write a history of science, in any case, because it is not like a biography or the history of a nation or of an art form, in which truthfulness is adequately served simply by sticking to the record. The heart of science beats in its ideas, not in events, but these ideas result in technological advances such as the telescope, or the battery, or the camera, which are intimately involved with economics, in turn a matter of politics, so that scientific ideas are really an expression of man's whole nature at a particular time and place.

Thus a history of science is necessarily a chronology of scientists involved with the daily life of their cities or countries. And yet, one insists, there must be an evolution of these ideas that can be discussed apart from the people who entertained them. But in every age genius is itself surprised by new ideas which were cued by discoveries and, uncomfortable as it may make us to point this out, by what seem to be wild guesses, stubbornness, dreams. There does not appear to be any logic to the growth of knowledge, and one of the great mysteries of history is the decay of civilizations and the eclipse of knowledge. Then, historical details about which no one argues do orient themselves into clear patterns for us, but into quite different, though just as clear, patterns for our fathers, and still other patterns for our children. Each pattern seems to have its own inescapably logical development. At best, history is a point of view.

Hence, of the countless theories of history, none has been able to establish a backward perspective that can be used to predict the future or that can even explain the present as satisfactorily as it explains the past from which it was derived. Furthermore, in a truthful history of science the profound errors and the violent wars of opinion, the wrongheadedness and wishful thinking, must also be included. The reader would have to learn the details of the ignorance that created confusion in a science before some theory resolved the confusion. A permanent confusion exists in every science at all times—including today—concerning that area of knowledge which has been freshly garnered by research and discovery and for which no adequate theory yet exists. To follow this shifting frontier is not our purpose.

We want to explain the present working view of matter, including some of the paradoxes with which contemporary scientists live in the firm belief that time will clarify them. We begin with the basic definition of matter.

Today's definition of matter

Matter is that which occupies space and has mass and energy.

The meaning of these few terms has evolved slowly during the past two centuries, and the definition cannot be attributed to any one person. The theory of relativity scrutinizes the terms with dismaying results, but for all practical purposes this definition remains the key abstraction of modern science, and we refer to it as such.

Matter occupies space. This distinction between space and matter is a categorical assertion that there is such a thing as empty space and such a thing as matter, and that the latter is clearly distinguishable from the former by its presence in it. In other words, I do not merely *imagine* that there is a tangible world floating in emptiness; there really *is* one, and wherever matter exists there is not nothing, because matter occupies space.

Philosophers since Aristotle have debated the distinction between the actual reality of matter and the sensory impressions of matter that reach the human mind, between an event and an observation of that event. In science the debate is not allowed, since it is irrelevant to the purpose of science. My senses, says the scientist as he sets up apparatus for an experiment, report to my mind impressions which are a truthful reflection of the world as it really is. The objects I handle, whatever they are composed of, occupy space, and their bulkiness, their volume, can be measured in three dimensions: length, width, and depth.

The metric system, with its meter and gram and liter, is the only one used by scientists to calculate volume, and we have something to say about it later in this chapter.

Matter has mass. A mysterious phenomenon of the universe is that pieces of matter attract one another. This attraction—there are several other kinds, as we shall see—is called gravity, or gravitational attraction. Weight is a measure of the gravitational pull between our earth and the object being weighed, and this pull is greater, the closer the object is to the earth's center of gravity. According to Newton's law of gravitation, if the distance between two lumps of matter is doubled, then the attraction between them is reduced to one-quarter of the original value. An object that has been carried to a mountaintop weighs less than it did at sea level; that is, it is pulled less strongly downward by the earth because it is farther away from the center of the earth's gravity. Far out in space the earth's pull becomes insignificant and the same object becomes weightless, but it is still the same object, it still contains the same amount of matter. Is its weight now zero? Yes. The gravitational attraction of other planets and of the stars beyond upon one another creates their great elliptical circlings through space, and we have to imagine that no object anywhere in the universe is free of the total gravitational pull of the universe at that particular spot. In advanced physics there is a different way of saying and seeing this. But what counts is that the weight or gravitational pull on an object varies according to its position relative to other objects in the universe. Thus the term "weight" is useless unless we specify where we are weighing an object.

The word "mass" was therefore proposed to describe the actual quantity of matter in an object that is unchanged by virtue of its position in space: the mass and not the weight of an object is always the same anywhere in the universe, except under conditions too extreme to consider here. Mass is a measure of inertia, which is the resistance of matter to being set in motion if it is at rest, and its resistance to change in its speed and direction if it is in motion.

To illustrate the difference between mass and weight, imagine a fish that weighs two pounds. Both the fish and the standard brass weight on which is engraved "2 lbs." are pulled with equal force toward the earth when they are placed on opposite pans of a balance. They will balance each other exactly. At sea level a spring-scale of the kind with which fish are weighed will read "2 lbs." if either the brass lump or the fish is hung on it, because the spring inside the casing has been calibrated and its indicator points to "2 lbs." when anything that weighs two pounds is hung on it at sea level. On top of Mount Everest the spring-scale's pointer reads less than two pounds for either the fish or the brass lump, which, however, will still balance each other on a pan scale. It is not their shape, nor their volume, nor their chemical composition, but the total amount of matter pulled by the earth that counts. On the moon they will still balance each other, for they will be pulled equally by the gravitational field of the moon. But the moon's field, which is only one-sixth that of the earth, will pull each one only enough for the spring-scale to register a few ounces instead of two pounds.

Thus, weight varies with an object's position in a gravitational field, while the concept of mass does not. In all the sciences, weight is measured by the metric system's units of gram and kilogram.

Matter has energy. Energy is a concept even more difficult to formulate than mass, because it cannot be grasped intuitively by the senses, as can space and mass. Energy is that which produces a force that acts upon matter. Thus, energy can be of many kinds, each of which can be transformed into another. For instance, gunpowder in a shell has a potential chemical energy that, the instant the powder explodes, is turned into the movement of hot gases which, in turn, transfer their energy of movement to the bullet, which then starts moving down the barrel of the gun. Some of the bullet's energy of motion is transformed into heat energy due to friction within the barrel. On the way to the target it loses another portion of its energy of movement that becomes heat energy in the molecules of air with which it has come in contact. When it hits the target, it is stopped and its remaining energy of motion is tranformed into movement of the target, which jumps.

When the chemicals in a battery begin to react and a current flows from the battery, chemical energy has been transformed into electrical energy. In a light bulb, electrical energy becomes heat and light energy.

There are also mechanical, radiant, and other kinds of energy. Of course these categories became definable only after the concept of energy had become accepted as having scientific justification and usefulness. Subsequently, techniques were worked out to measure each kind of energy by the amount of work solid substances could accomplish when associated with such energy. Thus, each kind of energy can be expressed by a formula which can be equated with work done; for instance, the lifting of a weight in a certain amount of time. Different types of apparatus can be rigged to make mechanical energy and electrical energy and heat energy lift the same weight, and so the amounts of different kinds of energy needed to do the same work can be compared. Each kind of energy has its own system of measurement, but these different systems can be translated into one another, just as pounds can be translated into kilograms. The chemical energy of a battery, measured in terms of some reaction potential, equals a certain amount of electrical energy in watts, which equals a certain amount of heat energy in calories, or of light energy in candlepower, or of mechanical energy in foot pounds. The words "force" and "work" also represent abstract concepts which had to be invented in order to discuss energy. Energy produces a force that does work. Here we do not bother with work and we restrict ourselves to a few kinds of energy.

Energy of motion is the most familiar to us, and it is called kinetic energy. A flying robin, a falling stone, a car traveling at 8o miles per hour on the highway or inching away from a garage, a raindrop sliding along a windowpane, all these objects have kinetic energy simply because they are moving. The heavier the object and the faster it is moving, the greater is its kinetic energy. A ping-pong ball hit hard enough has as much kinetic energy as a slowly moving golf ball.

The kinetic energy of any object is calculated by the formula $KE = \frac{1}{2} \times m \times v^2$, in which m is the mass and v is the speed, or velocity, of the object. If the velocity is increased, obviously the kinetic energy increases. Also, if a small m is multiplied by a large v, the resulting KE can equal a large m multiplied by a small v. The kinetic energy of any object can be compared to the kinetic energy of any other object mathematically, which is to say, quantitatively.

How was the formula derived? It is a relationship between mass and velocity that was learned from experiments carefully set up to find the relationship. When the velocity of an object is increased, the impact which the object would make on another is increased in a fixed ratio, and the formula expresses this ratio.

The basic stuff of the universe

All these ideas about matter developed slowly through the centuries and are such triumphs of the imagination that it is impossible to overemphasize their importance to our modern thinking.

The theory of relativity produces equations that relate the volume, mass, and energy of a piece of matter to the passage of time in a way that shatters the evidence of our senses concerning these terms, but they remain true attributes of matter all the same and are just as useful as they ever were. The formula for kinetic energy has to be modified under certain special conditions, but the terms and the concepts are as indispensable in the most advanced atomic physics as they are in the grocery store.

Reflection will stress that we have reduced the infinite variety of substances to a concept of basic stuff, matter—something which, whatever its special qualities and properties, can be discussed in terms of volume, mass, and energy. These are purely measuring concepts with which to compare all objects. There is no basic mass or volume. But during this century the concept of energy has had to be radically revised. We now know that what was once thought to be a property of matter is in reality a piece of stuff itself. Energy has become a thing in itself. There exist actual particles of pure energy and they are not at all slivers of matter but they do have mass. At the same time, the relationship between matter and energy has been abstracted into a mathematical formula. The stuff of the universe can exist either as matter or as energy.

This does not concern us at our introductory level, and it never concerns pure chemistry even at the highest level of research. It does concern nuclear physics and astronomy, but although the distinctions between physics and chemistry are fast vanishing, our original definitions of volume, mass, and energy will always hold in chemical work.

Standards of measurement

There was no such thing as a one-pound weight before someone cast a piece of brass and said, "Let this be called a one-pound weight and let all other one-pound weights in our fair city be balanced against this one, which we will keep in the Town Hall." "Let the length of my arm," said the king, "be called one yard, and let all the yardsticks in the kingdom be measured against my arm, so that they all will be the same length." Today our measuring instruments are matched by their manufacturers against official standards, and thus we really compare everything we measure to that standard.

The standard for our yardstick is not the king's arm but a platinum alloy bar sealed away from the environment in a case in Washington. The standard for the meter is a metal bar in Paris, and all the instrument makers in the world must cut their scales to conform exactly with that Paris standard.

In this way any scientist in any country knows that the centimeter marks on the apparatus he has bought from the local instrument maker, or from an importer, will conform exactly to the centimeter marks on every other instrument made anywhere in the world. He can publish his report with the conviction that if another scientist wanted to duplicate the experiment he would do so with identical scale.

All scientists of all nations use the metric system exclusively, and the reader should study its tables until he realizes the advantages of its decimal notation over the inch and foot, pound and ton, quart and pint units. The metric system is merely one more among the countless systems of measurement invented by merchants and craftsmen and artists and natural scientists and even philosophers, but it was put together expressly for modern scientific use and so it ties up linear measurement with volumetric and gravimetric measurement. One liter, about a quart, is equal in volume to 1000 cubic centimeters, and 1 cubic centimeter of water weighs 1 gram, at 4 degrees centigrade. The ultimate standard for the whole metric system is the meter, which was originally conceived as a calculated fraction of the earth's circumference. The earth thus replaced the king's arm. Errors have been discovered in the standards the French Revolution's scientists set up with this system in their passion to create a rational society, but the errors do not interfere with ordinary laboratory accuracies.

Another standard without which there could be no science at all is the thermometer, which is simply a sealed glass tube with a little mercury at the bottom. As the temperature around the glass tube rises and the glass gets warmer, the mercury is heated and expands to fill more of the tube. If we put an unmarked thermometer into boiling water we can mark on the glass the point to which the mercury expands. If we put the thermometer into ice water we can mark on the glass the point to which the mercury shrinks. If we mark the freezing point of water 0 and the boiling point 100, and divide the distance on the tube between them into a hundred lines, we have a centigrade thermometer, first designed by Celsius. If, instead of 0 and 100, we write 32 and 212, then we have a Fahrenheit thermometer.

The standard for time has been the solar year or the sidereal year, the latter used in astronomy because it is a shade more accurate. But these have become unsatisfactory in nuclear research, and a new standard has been set up: the frequency of the vibrations of cesium atoms or ammonia molecules, or of the electromagnetic waves emitted by other atoms.

3

The Scientific Approach to Search and Research

Assumptions concerning the universe as a whole

THE notion that matter can be discussed fully in terms of mass, volume, and energy means that apart from the special properties of a particular kind of substance—leaf, sand, flesh—all matter obeys universal laws. There is, in other words, a universal nature to matter even though any one piece of it is utterly different from another piece. The Greeks had this notion too, but they could discover only a philosophical way of discussing it.

An integral part of our concept is the laws of conservation of mass and of energy, which state in their modern version that the sum total of all matter and all energy in the universe does not change, no matter what transformations they undergo. Accurate measurement can never be the road to knowledge unless it is assumed that matter cannot be spirited up out of nothing or vaporized out of existence, and unless it is assumed that energy cannot be collected out of nothing at all or made to vanish without a trace. When men believed in the spontaneous generation of blowflies on rotting meat, the most brilliant reasoning could not make a very important matter out of weights and measures.

The success with which controlled experiments began to gather knowledge that could be verified by anyone using the same principles of control—accurate measurement—gradually dispelled belief in *extra*ordinary events in the material world. Rigorous standards in what we call the quantitative approach to nature assembled information that would have been forever beyond the power of pure reason to discover. By the time the laws of conservation were actually spelled out, everyone knew they were true, or at least that we had to assume they were true. But they had to be formally stated because, after all, though it cannot be proved that the universe was created once and for all, with all the matter and energy it now contains present at the beginning and destined to be present at its end, our measuring passion is sustained by an absolute belief that the laws of conservation of mass and of energy constitute a principle of nature.*

There are other assumptions necessary before the mind can take the scientific approach seriously, and these may seem full of common sense to the layman but they have tortured philosophers. They are not formulated as laws or theories or gathered into a credo.

For instance, science assumes that the universe is an orderly structure built on the same principles here on earth as in the farthest reaches of space. There is no ideal corner, or un-universe-like area, in this universe. Its laws are uniformly in force from one end to the other and from start to finish.

The primary principle of this order, scientifically speaking, is causality, meaning that every event has its cause or origin in a previous event. There are no Acts of God, no gratuitous accidents, no beginnings without antecedents.

Although the concept of natural law is also being re-examined, the usefulness of considering natural law a kind of unbreakable rule of behavior, a kind of statement concerning that which actually is, rather than an ideal law in the social sense which can be broken—this will not change for some time to come either. Briefly, the doubt as to the implacable nature of natural law came in with the principle of indeterminacy, which elevated the statistical as-

---

* Recent cosmological speculation has produced the theory of the steady-state universe, which postulates the continuous creation of matter in space, thus appearing to violate the conservation laws. But the predictions of the steady-state theory have not yet been tested by sufficient observation, and at the moment the question of whether or not matter and energy are still being created in outer space does not concern the practicing scientist, nor, as far as one can see, will the laws of conservation ever become inapplicable or untrue or useless here on earth.

pect of events to the distinction of a law. It need not concern us at the moment.

It is taken for granted that man's reasoning faculties can be focused upon this universe through the impressions it makes upon him, and that he can pry out of the events he observes their cause and their effect. Thus, one by one, he can discover cycles, statistical realities, and properties, and can formulate mathematically precise relationships—laws. With these laws the future can be predicted, because they will continue to be operative.

The final assumption, or perhaps it is the first, is that it is good for man to analyze the universe scientifically.

The rule of science

Every one of these grand assumptions is being questioned today, but scientists are not bothered in the least. Their work is guided by the scientific method, and even if all the assumptions we have listed, and more, turn out to be false, the scientific method will remain unblemished and just as powerful as ever.

It was first defined by Francis Bacon in the early seventeenth century as a system of logic totally different in its aim and its procedures from that used in theological or philosophical arguments. But before the method was accepted in scientific work generally, all the assumptions we have mentioned had to gather significance against the assumptions of Aristotle and of Christianity. Not that the scientific method opposed Christianity. It simply assumed that the search for truth about nature could be conducted independently of the search for ethics, independently of the search for the meaning of existence, and quite apart from the search for God. The search for God's will was not the aim of the scientific method. Bacon's system of logic sought to uncover the realities of nature.

Today many scientists as well as philosophers and psychologists claim that the scientific method is not at all a definable process of thinking and that some of the nineteenth century's faith in it is based on a false distinction between inductive and deductive reasoning. The great advances in knowledge are really incomprehensible leaps of recognition in the minds of brilliant men who grope intuitively—what we call intuitively—for such relationships. Intuition cannot be regulated. In any case, whatever the proper definition for it may be, what is called the scientific method has worked with stupendous success during these last two centuries and is working with undiminished vitality today.

The scientific method is at the very least a guiding principle in research and ordinary laboratory work and it certainly guards the mind against the longing to select from its impressions only those that will bolster some private dream and to discard impressions that oppose the dream or are irrelevant. Thus, one of its main uses is in training students to acquire the objectivity necessary for a scientific point of view.

Practically speaking, the scientific method results in a system of manual procedure, of instrument reading, and of mathematical calculation which attempts to relate data about matter and energy in a particular situation, without regard to values the situation may have on the human scene. Its classic conception can be described more or less as follows.

A situation, or an event, in the material world, such as a thunderstorm or the germination of seeds or a platypus eating or a falling stone, is observed from as many approaches as possible until sufficient data have been collected to suggest a hypothesis concerning the cause of the event, or its structure, or its link with other events. Experiments are invented to test the hypothesis over and over. If it is a sound one, then not only does it stand up to rigorous testing but it gradually becomes applicable to more and more facts until it uncovers all sorts of hitherto unsuspected relationships among other events and is indispensable to thinking about a variety of phenomena. Eventually the hypothesis acquires the fullness and the succinctness of a theory.

But all this is not as simple as it sounds. Theories are not acts of Congress enunciated to solve certain difficulties. The function of laws is not at all clear, either. Some laws were discovered by researchers hunting for what they were certain existed as a mathematical relationship in a set of factors. Some laws were stumbled upon by chance without anyone's having dreamed of their existence. Still other laws originally were simply useful guesses that only later were justified by hard work. It is a commonplace in science to have theories for which no man can claim credit, and many theories have grown out of an attempt to explain formal laws after their discovery, but the reverse has happened too. The point is that a great deal of experimentation is *ad hoc* and has no formal hypothesis behind it, and yet it is scientific because the scientist abides by the essential meaning of the scientific method.

One of the most important aspects of scientific work is that both laws and theories can be amended and even discarded after long and honorable usage, during which they were considered impeccably "true." The humanists have traditionally considered this a great weakness in the scientific search for truths—what is scientifically true one year may very well be only partially true next year—but this does not hamper scientists one little bit. In fact it gives them confidence that the system has self-criticism built into it and that no mere illusion of truth can usurp their minds for long before someone discovers the error.

In any case, every attempt to apply the scientific method without full use of the imagination, of intuition, of all available knowledge, of the undefinables of past experience, of common sense, as well as of a refusal to be commonsensical at the right moment, and of a willingness to gamble and a willingness to repeat endlessly, leads to painfully unrewarding results. The most definite thing one can say about the scientific method is that its essence is the empirical point of view and that this point of view is most prominently displayed in laboratory research.

Empiricism states that all knowledge is derived exclusively from experience and that even reasoning is a learned process. Logical extensions of this belief lead to disturbing concepts of "being," but it dominates every realm of activity in our culture today, including some of our religions. Certainly the departments of psychology, sociology, history, economics, anthropology, even the fine arts, have been trying for several generations to establish a scientific orientation toward human values, and in certain respects they have succeeded, whereas in other respects they have not.

But let us try to imagine a genuinely earnest scientist hard at work.

Shades of Gulliver

Our model scientist comes from the Island of Laputa, to which Gulliver paid a visit and which he described in scurrilous terms. Scientists were fools and worse, in their passion to measure and analyze, was Gulliver's conclusion. A Laputan visiting us would at once be struck by the vast number of cars—about which he has never heard—streaming along the roads, stuck in traffic jams, parked along the curbs, stored in garages, and heaped in car lots. Not only their numbers but their speed and precision of movement fascinate him. Knowing next to nothing about our society, he decides that cars are a useful object of study and he gets a grant from the University of Laputa to extend his visit.

In a very short time he will formulate the tentative hypothesis that a car starts to move after a person gets behind the wheel but it never moves unless someone is behind the wheel. If he now proposed that a car is for transporting man from place to place, he would not be a good scientist because all sorts of other things may be happening to the man behind the wheel. The point is to find out what a car is.

He notes that it does not run until the man has turned a key, when a noise in front and a noise in back become audible; from time to time, gasoline is poured into the rear and oil into the front. First it was the man who seemed to make the car run, then it was the key, then it was the noise front and back, and now the gasoline and oil, back and front.

He measures the distances, speeds, and places cars go between refills, but no patterns are revealed. He details a graduate assistant to study the amounts and varieties of fuel, but again no significant pattern can be found.

The scientist now moves to the next level of observation and attacks with a sledgehammer as many cars as he can capture. Sometimes a single blow prevents the car from running; at other times he practically demolishes a car and still it runs. Gradually he brings the focus of his attention to the hood, having worked over everything else carefully, but just then he dies.

His assistant decides to continue the great work that once excited professional circles in Laputa, because he cannot bear the thought of starting all over again on something else for his doctorate. He sends for heavier sledgehammers, and then, quite by chance, he has the brilliant idea of using a crowbar instead.

In the first heat of excitement over this brilliance, he logically pries off the four wheels. The resulting depression brings him to the brink of giving up the whole business and simply publishing the dead professor's notes, for which he might get a few credits.

But in the process of deciphering the professor's spidery script, he suddenly has a hunch: how would it be to pry off the hood, which was the professor's final enthusiasm? So he pries off the hood, and there it is—a totally new and unsuspected world that he names "engine." A series of swift experiments proves that the engine is the source of power that turns the wheels.

But what goes on inside the engine? How do gasoline and oil become the spinning of wheels? He begins to take apart the engine, to analyze it with both sledgehammer and crowbar and, from the bits and pieces that result, he assembles a diagram of its parts in relationship to one another that is quite wrong but that has a vague resemblance to an engine and will serve to design the next series of experiments.

One can hazard a guess that four or five generations of scientists in teams, using the scientific method with strict discipline, will collect almost as much information as a teenager can be taught in one day by a good mechanic.

But imagine the dizzy joy with which our Laputan discovers that the noise in the rear is really the noise of the engine piped to the rear through a tube which also carries the fumes! Yet he cannot understand—and never will—why there is such an exhaust pipe and muffler, unless he already knows that our society likes quiet machinery. Unless he knows that we do not like mud flying through the air, he will never understand why there are mud guards. As for the variety of styles in car shapes, what on earth can he make of that unless he understands fashion, the profit motive, status symbols, and all the rest? Now that he knows what a car is and what it does, as a car, in order to find out why people invented cars and how they use them, he must become a student of human values. And even then the original impact of the traffic-clogged streets and highways continues to retain a quality of mystery—just as it does to us.

The essence of the scientific method as defined in any textbook is accurate observation of nature, which means precision measurement, a fact we did not stress in our fantasy because it would have slowed up the argument. A proper Laputan would analyze the paint job that flings chips in his face when the sledge hammer hits it. He would analyze chemically and take photographs of every bolt. In real laboratories the situation is never so ludicrous, mostly because all research is related to a vast body of information already compiled and available, but partly because we cannot see ourselves quite so clearly as we can imagine the Laputan.

Knowledge is ordered information

Let us re-examine our categories of properties and forces associated with objects.

Aristotle was the first man in the West who systematically reduced the uncountable number of things to families whose members were related because of some palpable similarity. His system was based on careful observation, not on wishful thinking or on hearsay or on traditional beliefs. For instance, a butterfly whose wing markings mimic a leaf is classified with butterflies and not with leaves. In thinking about butterflies in general we must include in our abstract model of butterfly the potential that some may have wing markings that mimic leaves. We will never think that leaves mimic the wing markings of butterflies, but why? Because, said Aristotle, butterflies are on a higher level of existence, according to the criterion of independent mobility and of other criteria he had established. There are difficulties: is a sponge an animal or a vegetable? Without a microscope he couldn't tell.

Aristotle's classification, though it was intended to illustrate a moral order with man at the top and the whole revealing a divine purpose, was amazingly modern. But the modern classification, though it arrives at much the same order as Aristotle's, has no moral significance. The greater complexity of a bird does not make it a finer or more beautiful thing than a toad. The theory of evolution denies that there was an original ancestor for each and every living form, and denies that each form breeds true to itself unchanged to the end of time, and denies an order of things. The beasts are not one thing, soulless, and man another, endowed with soul. They and we have common origins.

The properties of objects by which they are recognized fall into two categories, the physical and the chemical. Physical properties include density, hardness, color, solubility, magnetic and electrical behavior, boiling and melting points, the angles at which light is refracted, crystal shape, surface tensions, and many other aspects. Each of these physical properties can be expressed as a number on some appropriate scale, designed for the purpose, that compares what is being measured to a standard. We have already shown how temperature on the centigrade thermometer is expressed as a number against the standard zero degree of the temperature of ice water. Hardness is measured by a scale of ten substances that gives talc, the softest of the ten, a hardness of 1, and diamond, the hardest, a hardness of 10. An unknown material is scratched by the standard substances until it is determined that its hardness lies between two of the ten—it is harder than one but softer than the next one in the scale. Hardness is therefore reported as a number—for instance, 7.5. This means that the substance is halfway in hardness between quartz and topaz.

Chemical properties, on the other hand, cannot be turned into comparative figures. They concern the actual chemical reactions in which a substance will participate. The temperature at which a substance bursts into flame, called the kindling temperature, is a physical property; but the ability to burn with a flame, that is a chemical property.

As for the category of energies that the ancients believed to be spirits controlling matter from beyond the material world, science has transformed them into laws. The laws of electricity, magnetism, light, radiation, waves, gravity, torsion, mechanics, are all mathematical relationships—not numbers, but formulas and equations using numbers derived from observations. The gods of myth have become the immutable laws of science.

And what of the inescapable process of the mind—as old as thought—that invents models of the cosmos? For two thousand years western man could not imagine a universe of which he was not the most significant part—the center. In the nineteenth century scientists became convinced that the universe ran like a marvelous clock and that even man's passions, however unlikely it seemed, were necessarily wheels in the clockwork. This was not felt to be a fatalistic view for the simple reason that it was scientifically marvelous and irrefutable. Then the mid-twentieth century created a billowing, twisting continuum in which man is just barely, if at all, distinguishable from a flux of forces.

In the following chapter we begin to examine the framework that supports the present model, in which chemical reaction is seen as an activity of subatomic particles, of fields and of positive and negative charges—a flux of forces.

4

A Theory of Basic Particles Explains Heat

The kinetic molecular theory

THE first question we ask about matter after having managed to separate it from space was asked in Greece by Democritus: is matter continuous or discontinuous? His atomic hypothesis was attacked as being atheistic, and fifteen hundred years later Dante put him into the pit of hell. Even today it is impossible to explain why there is no chaos if ultimate pieces of matter merely attract or repel one another. If I consist only of unpurposeful, eternal atoms, what in me controls them? It seems impossible that each atom has conscious will, and just as impossible that each atom has within it a tape directing its destiny to the end of time. Yet atoms cooperate with one another to create sand, leaves, flesh, the eyes that read these lines. If the universe has a purpose, that purpose must be obeyed by the least particle in it. Endless questions occur to any honest mind, today or two thousand years ago, when it meets the idea of atoms for the first time.

Democritus's atomic hypothesis complicated rather than simplified what seemed to be the evidence of the senses, even if it did brush aside spiritual debate and focus exclusively on the material aspect of nature. But the assumption that atoms were too small to be seen made them the subject of debate rather than of research which could prove or disprove their existence until the scientific view became established and, following Lavoisier's leap out of two thousand dreamy years into the hard, gravimetric realities of chemical action, the idea began to gather substance, gradually and confusingly, that matter *had* to be discontinuous. Finally it became imperative that this be stated clearly and formally as a theory, not against any rival theory of continuous matter, but because certain other assumptions about matter were clamoring to be made also.

The cumulative result of various intellectual needs was the kinetic molecular theory, which cannot be credited to any one man and for which no publication date can be given. It is the final distillation of proposals that scientists put forward here and there, over a period of time, in more and more useful formulation until it was in general acceptance. By all odds it is one of the most important scientific theories concerning the nature of matter, for without it there could not be any science in the modern sense.

The easiest way to introduce many theories is to make a series of flat statements without qualification or argument and when the whole is before the mind's eye, then to examine the details. It is the procedure we shall follow now.

All matter consists of particles that are too small to be seen even under the most powerful optical microscope. These particles have all the properties of the ordinary bulk matter which they compose, since they are merely the smallest particles of it that can exist separately. Each particle of a specific substance has a mass, volume, and energy, proportional to its tiny size, and in every other way is identical with all other particles of that substance.

The particles are in constant rapid motion. This motion is an inherent aspect of these invisibly small but real bits of matter. The ordinary law of kinetic energy applies to this movement: the kinetic energy of any particle is $\frac{1}{2}mv^2$, just as though it were a cannonball or a star. The kinetic energy of a particle is an expression of heat energy.

Heat, like light, has been one of the great mysteries of nature from antiquity to this day. The kinetic molecular theory states flatly that the motion of ultimate particles is equatable with heat. The speed with which a particle moves is an expression of the amount of heat it has at that instant. The faster it moves, the more heat it has.

The particles move in a straight line until they collide with another bit of matter; then they bounce away in some other direction but always in a straight line, and without losing any of their energy to friction in the collision. The particles have perfect resilience. However, they will change their kinetic energies in the familiar way of all bulk matter, as, for instance, do billiard balls. A slow-moving particle hit from behind by a fast one is speeded up, while the fast one is slowed down, but the sum total of their kinetic energies remains the same. In the world of bulk matter, perfect elasticity is unknown, as there is friction between surfaces. Two billiard balls when they collide will change each other's speed and direction of motion, and they will also roll to a stop in a short time. The ultimate particles of matter lose not a whit of their energies in collisions; they simply exchange speeds. If two particles collide, their total heat before and after is the same, but the originally slower particle after the collision is traveling faster and is therefore hotter, while the formerly speedier particle is now cooler and moving more slowly than it was. Heat and molecular motion, according to the theory, are two ways of speaking about the same thing.

This equating of heat and motion works so wonderfully well in our technology and science that, when the mind gets used to it, it seems the most natural thing in the world to see things that way. Furthermore there is abundant evidence that the theory does state something very close to the truth about this aspect of matter.

We shall call these basic particles of matter molecules but with a reservation. In Chapter 6 the term "molecule" is redefined and given a much more precise meaning. Here molecule means simply any basic particle, whereas after Chapter 5 it means only one kind.

What everyday phenomena does the kinetic molecular theory explain?

If you step into a house where cabbage is cooking several doors down the hall, you will smell it. Is it possible that the cabbage attenuated itself through doors, like some kind of ectoplasmic gum? Or have actual pieces of actual cabbage traveled through small crevices and through the air to the front door? If a girl opens the door her perfume will mingle with the smell of the cabbage. A drop of ink in a glass of water spreads out until the water is uniformly colored. Sugar, when dissolved in tea, seems to vanish, yet every teaspoon of the liquid is uniformly sweetened.

These are examples of diffusion of one substance into another. The particles of cabbage diffused through the particles of air is a phenomenon that cannot be explained if we assume that matter is continuous. We also note that the hotter the room, the faster the particles of cabbage and perfume and the aroma of the roses on the mantelpiece will diffuse—the faster all molecules, including those of the air, will move, resulting in more collisions each second. The molecules of ink, tumbled about by the molecules of water which are sliding about, are hammered through collisions into uniform distribution, faster in hot than in cold water. Sugar molecules dissolve and sweeten a cup of tea faster if the water molecules are hot.

Another phenomenon that cannot be explained except by the kinetic molecular theory is the so-called Brownian movement. When microscopically small dust is dropped into water, the tiny motes can be seen in a microscope to be dancing in all directions, in shorter and longer darts, and at random angles. The movement is incessant and chaotic. The same dust will lie absolutely still on top of the table. The kinetic molecular theory explains that the motes of dust are enormous compared to molecules of water and have lost their kinetic energies, but the fastest-moving water molecules slam the motes hard enough to bounce them a short distance through the mass of much more slowly moving water molecules. As there are not many of these extremely high-speed water molecules, their bombardment is haphazard and intermittent, but they are always around and so the movement does not stop.

The gas laws

Science is full of models put together out of verifiable data in mathematical relationship, out of theory, axiom, and metaphor, and out of sheer poetic visions. Their purpose is to enable us to discuss abstract truths, or perspectives, or generalities, and the fact that they cannot be transformed into real models makes them more useful than wood, metal, leather, or plastic could be. As intellectual constructions they escape the limitations of brute matter and can be isolated in the imagination from the rest of the universe and studied with an exclusiveness impossible in reality.

The first such model of matter that we examine here is built with the kinetic molecular theory. Imagine matter in a state when each particle is totally free to do as it pleases within the limits of its nature and its environment. A gaseous condition seems to be the one in which molecules are most free. Therefore let us imagine a certain number of gas molecules in a container with a movable top, such as a cylinder with a piston that can be pushed up and down, and with its parts fitted so perfectly that no gas can escape and no air can get in. No matter how many molecules of the gas there are in it, no matter how large the container, the molecules will distribute themselves throughout the whole container because of their constant motion, which makes them pound one another in ceaseless collisions that push them apart in all directions.

Inside the container a molecule is not pulled in any one direction more than in any other. What scientists call chaos will reign, for there will be no orderly pattern of movement anywhere within the container. There will be no streaming of a lot of gas molecules all in one direction. It is true that gravity will add a slight speed to the downward-moving molecules and reduce slightly the speed of the upward-moving ones, but gravity will not orientate the actual direction of movement. Thus a sample from the top of the container, as long as it isn't, say, a mile high, will contain just as many molecules per cubic centimeter as a sample from the bottom or the middle, and they will be hitting one another and the walls of the container and the face of the piston to produce with their impacts what we think of as gas pressure.

If now the piston is pushed in so that the space available to the molecules is only half what it was at the start, then the distance between each molecule is only half as much as before; and if the temperature has been kept the same so that the average speed of each molecule has not changed, then each molecule in its totally random bouncing will hit something twice as often per second. The pressure which the gas as a whole exerts on each square centimeter, through the impacts of the same number of molecules, will be twice as much. Since the bombardment on the face of the piston will be twice as much, it will take twice as much pressure to keep the piston in.

Conversely, if the piston is pulled out to enlarge the amount of space available to the molecules inside the container to twice what it was, each molecule will travel on the average twice as far without hitting anything. This means that all the molecules collide only half as often per second as before. The pounding on the piston will therefore be only half as much on each square centimeter, and the pressure, which is really a measure of these impacts, will be reduced to half what it was on the whole piston. Again, this is true only if the temperature has not changed.

These statements can be tested with an actual cylinder and a sliding piston hooked up to pulleys and weights, and experimenters have proved over and over that the statements are correct. Furthermore, the relationship of pressure to volume is mathematical, just as our intellectual model supposed it would be, and it is called Boyle's Law, after the British chemist Robert Boyle, who discovered it in the seventeenth century. The law states that the pressure of a fixed weight of a gas upon the walls of its container, and on any point within it, is inversely proportional to the volume of space the gas occupies, if the temperature remains constant. (Inverse proportion means that as the amount of pressure increases, the volume decreases; whichever gets larger, the other gets smaller. The proportion between them is an inverse one.)

To know this about a gas is of inestimable importance to science and industry because we live in a sea of gases, our atmosphere, which presses upon us and on our laboratory equipment and factory installations with a ceaseless pounding of gas molecules. Many of our engineering problems in industry concern the handling of gases hot and cold, explosive, corroding.

Now imagine our container, with its movable piston, fitted out with several thermometers in different places. They will all read the same, just as one would expect. But will all the molecules have that particular temperature? Even if we imagine an instant when all the molecules have the same kinetic energy and therefore the same heat, the next second some will be traveling faster and others more slowly, owing to their angles of collisions, and their individual kinetic energies, thus their heat, will be different.

We have avoided using the word "temperature" loosely. It is correctly used only with a large quantity of molecules, and when we are speaking of a single molecule we can only speak of its kinetic energy, or heat.

If the temperature of the container's walls is kept steady by appropriate shielding, each molecule that hits the walls will either lose or gain heat—speed, kinetic energy—to conform more nearly with the kinetic energy of whatever molecule of the wall it hit. Each molecule in the gas will keep adjusting itself to an average of all the various speeds, which will be equivalent to the average temperature of the walls. The thermometers will not be able to register the high or low heat of a single molecular impact, but only the average of them all.

Though there will always be some molecules whizzing at lightning speed and others at a complete standstill at the instant of a head-on collision, no molecule can retain for more than a fraction of a second a kinetic energy much different from the average. The slow ones will be hammered with extra vigor, and the fast ones will hammer with extra vigor, and both will be brought to the average quickly.

Even in a mixture of light-weight and heavy molecules, there will be a steady reading on the thermometers because, though light molecules move faster than heavy ones, the formula for kinetic energy shows that the impact of a light and fast particle will equal the impact of a heavy slow one.

In any gas or in any mixture of gases the kinetic energy of most molecules is close to the average, but it is theoretically possible that at some instant not a single molecule will have the exact average energy.

Suppose now that the walls of our container are heated from the outside. As a gas molecule hits the wall it absorbs heat and increases its speed and it will pound the next few molecules it hits harder, increasing their speeds. In a few seconds all the molecules will be moving faster, the number of impacts per second will increase, and the total pressure on the walls will increase. The piston will be forced out by the hotter gases and the volume the gases occupy will be increased—if the outside pressure on the piston is not changed.

Thus, we discover a direct relationship between volume and temperature, and it turns out to be mathematical, just as the volume-pressure relationship did, only now the proportion is a direct one. The relationship was discovered in the eighteenth century by the French physicist Jacques A. Charles, and his law states: the volume of a fixed weight of gas increases in direct proportion to the temperature, if the pressure is kept constant.

Charles's law is just as invaluable in research and industry, and in theoretical physics, as Boyle's law and in fact the two laws are usually used in a single formula. With that formula we can speculate upon the formation of stars out of cosmic gases, and we can also build a city gas tank so cleverly that the changes of pressure and temperature due to uneven use of the gas and the sun's effect are all safely taken care of.

Absolute zero

Now suppose we begin to cool our whirling, spinning, darting molecules of gas, which pound one another ceaselessly in all directions, and we measure the drop in temperature that we bring about —by cooling the container—and we also measure the amount that the gas shrinks, which is what Charles did. We find that if we start at $0°$ C. and lower the temperature one degree centigrade, then the volume of the gas shrinks $\frac{1}{273}$ of the total. No matter how large or small our container, the piston will sink in, as we lower the temperature from 0° C. to −1° C., one part out of two hundred and seventy-three parts of the total. Each degree that we cool the gas it will shrink this same fraction. If we cool it by 10 degrees, it will have shrunk $10/273$ of the original volume. If we cool the gas 273 degrees below 0° C., it will have shrunk $273/273$, which means that it will have vanished. There will be no volume of gas left in our container.

Obviously something has gone wrong with our reasoning.

What has been happening, did we say, as we cooled the gas? It shrank together. But why? Because the kinetic energy of each molecule was reduced. We can say that the kinetic energy of the gas molecules was reduced $1/273$ for each degree drop in temperature, starting at 0° C. Then, when we have cooled the gas to −273° C., there will be no more kinetic energy left in the molecules. All motion will have ceased.

Can such a temperature be reached? Not quite, but modern research has come to within a fraction of it, and the theoretical end is called absolute zero.

What happens in reality when we cool a gas? Everyday experience tells us that water vapor condenses into liquid, and that liquid water freezes into ice. And thus we can formulate one of the important categorizations of matter: every substance can exist in a gaseous, a liquid, or a solid state, or phase. The only difference between these states or phases is the temperature. How does the kinetic molecular theory explain the transformation of one state of matter into another?

Accounting for Three States of Matter

Gas to liquid and vice versa

INSTEAD of an ideal gas in a container whose temperature and volume we can regulate from the outside, imagine actual water molecules in the gaseous state, in a sealed container of fixed volume whose temperature we can regulate from the outside. The temperature of the container is above 100° C., which is the boiling point of water, and therefore all the molecules in the container are in the gaseous state.

Now cool the container. As the hot molecules hit the colder walls and lose some of their heat they bound away more slowly, and as the cooling continues the average speed of all the molecules decreases according to the relationship expressed by Charles's law. Eventually a few of the extra-slow molecules will be moving slowly enough so that when they collide they will linger in each other's vicinity and at that instant certain forces of attraction, of which we have had no inkling until now, suddenly exert themselves. The attraction between the molecules has been inoperative at high speeds, or high kinetic energies, but two slowly moving molecules will find themselves caught. They will cling to each other. And at that instant they will have given up a certain amount of energy which they have only in the gaseous state.

As the temperature of the container is lowered, more and more clusters of a few low-speed molecules will form. Each cluster exerts a combined attraction on more and more slowly moving molecules, and eventually the clusters will become large enough to be visible as a mist, and then as droplets. Within each drop the molecules continue to move, but they are trapped by their mutual attraction from leaping away as they used to in the gaseous state. They simply slide about, around, over, and under one another in random darts. A gas fills whatever container it is put in, but a liquid maintains a constant volume in whatever container it occupies.

The kinetic molecular theory states that the kinetic energy depends on heat energy, which can be measured as temperature. A thermometer in boiling water and a thermometer in the vapor just above the boiling surface will read the same: 100 degrees C. Therefore the average kinetic energy of the liquid molecules must be the same as the average kinetic energy of the gas molecules above it. An average molecule in the liquid state will be moving as fast as an average molecule in the gaseous state.

If we boil water in a kettle its temperature remains at 100 degrees C. no matter how fast we boil it, but we have to keep adding heat to keep it boiling. What happens to that heat? It is absorbed by the molecules as they escape their liquid state and become gas. Countless experiments have been conducted to discover how much heat is needed to pull apart liquid molecules caught in the liquid web of their attractions, and this quantity, which is called heat of vaporization, is different for every substance. It is expressed in calories per gram, and it is often used to identify a substance.

The heat of vaporization which a liquid molecule must absorb before it can become a gas molecule is released by it, when it cools again to liquid, as heat of condensation. The two terms are interchangeable because they are the same quantity of heat.

The point to remember is that during the transition from gas to liquid, or liquid to gas, the actual kinetic energy of the molecule does not change. It merely absorbs or releases its specific heat of vaporization, internally, as it were. That is, besides its kinetic energy, a molecule has other energies locked up inside it.

Now consider the surface of the liquid that has collected in our container. The molecules in the uppermost layer are being pulled downward by those deeper in the liquid and by those at the sides, but nothing except random molecules in the vapor above it pull upward at those liquid surface molecules. Thus a liquid is pulled inward by a net of clinging molecules wherever it is exposed, and this physical property, which is called the surface tension, can be measured and varies from substance to substance. The surface of water will support dust, leaves, beetles, but the surface of mercury will support nails.

Suppose a liquid molecule because of a series of collisions is moving so much faster than the average that when it hits the surface net it breaks right through and escapes into the gas state. Such a molecule must have collected its heat of vaporization before it could escape. Therefore each such movement carries with it a tiny amount of the total heat in the liquid, leaving it that much cooler.

A glass of water will always be a little cooler than the air of the room where it sits. Water in a clay jug seeps through the walls of the jug and forms a film on the outside, increasing many times the exposed surface of water and increasing vaporization; water inside the jug is robbed of heat and is always several degrees cooler than the air. Wet wash whipping on the laundry line is cooler than the wind. Liquids with low boiling points, such as alcohol or ether, chill the hand as the molecules pick up their heat of vaporization and become gas.

But consider again the liquid water in our sealed container. Fast molecules bursting into the space above the surface will increase until a few, because of collisions, are again slower than average, slow enough so that when they chance to hit the surface they are caught by the liquid and held fast. At that instant they give up their heat of condensation to the liquid. Thus the temperature of the liquid and of the vapor in a sealed container are the same; the molecules escaping from the liquid will equal the molecules returning to it because the pressure exerted by the vapor molecules in the space available to them will be constant, for any one temperature. Every liquid has its own vapor pressure, which is expressed as pounds per square inch, or grams per square centimeter, and this figure is used as an identifying property.

To put it another way, the total number of molecules above the liquid, in our sealed container, does not change if the temperature is kept constant. The total number depends on the amount of space available to the vapor, but it does not depend on the amount of liquid present. Molecules will be moving constantly in and out of the liquid, as their kinetic energies change, but the unit pressure they develop above the surface remains constant. The vapor pressure of a thimble full of sea water is identical with the vapor pressure of the ocean from which it was dipped.

The condition at the liquid surface in our sealed container, with the number of molecules leaving the surface equal to the number returning to it, is called a dynamic equilibrium between the liquid and gas phases.

Systems that achieve dynamic equilibrium abound in chemistry and physics and are one of the basic phenomena of nature. After the concept becomes familiar it is impossible to imagine a universe which is not composed of interrelated systems in dynamic equilibrium, and in fact even the life sciences have accepted the idea that every event, however large or small, is also part of a larger system in which the parts are woven into a dynamic balance.

Liquid to solid and vice versa

Let us return to our sealed container with the liquid and gas phases in equilibrium, because the temperature is being held steady. Now resume cooling this system, which is to say, put the container into a cooling environment, such as ice water. At once the vapor pressure drops. Fewer molecules will be able to maintain the necessary kinetic energy in the gaseous state. Within the liquid, the number of slowly moving molecules will increase. The viscosity of the water increases—that is, it will flow more sluggishly, the molecules sliding past one another more reluctantly. At last a few of them are moving so slowly that all at once a fresh force of attraction between them is felt by each and they become locked together. The next slowly moving molecule sliding past is pulled up against them. All through the cooling liquid the slower molecules begin to cling to one another, and at these points kernels of the solid state will grow until they become visible as crystals of ice. The tightly fused molecules continue to attach more molecules as the system is cooled, and eventually all the liquid is turned into solid.

If, at some point when only a little bit of solid is visible, the cooling is stopped and the temperature is held steady, the crystal will not grow or melt. If we add heat, it will melt. If we remove heat, it will grow.

Exactly the same sort of thing is happening that happened when gas molecules condensed. Each molecule in the liquid now gives up a specific amount of heat, the heat of fusion, as it becomes part of a crystal. To melt the solid, heat of fusion must be added to the molecules in the crystal.

The kinetic molecular theory does not say that molecules are in motion only in the liquid and gaseous phases. All matter is made of molecules in constant motion, down to absolute zero, where motion ceases. In the solid state, therefore, motion must still exist simply because the temperature is not absolute zero, or —273° C. In the solid, and most solids are crystalline, the molecules cannot roam about, for they are firmly locked in the embrace of their neighbors. But they vibrate in one spot. All the molecules of a solid are constantly vibrating with the same kinetic energy that causes the liquid molecules to dart about across the solid's surface.

And again, all the molecules in the solid are not at the same energy level and some at the surface will be hit by vibrations large enough to dislodge them and kick them out, together with their heat of fusion, into the liquid state. Slow liquid molecules, touching the solid surface, are trapped and give up their heat of fusion.

Thus a dynamic equilibrium is set up at the interface between solid and liquid. Any single molecule can at one instant be in its liquid phase, the next in its gaseous or its solid phase, depending on how much hotter or cooler it is than the average. A thermometer in the cake of ice, a thermometer in the liquid, and a thermometer in the vapor will all read the same temperature: 0° C. (For the sake of accuracy we must add: at normal atmospheric pressure). The average kinetic energy of every molecule in this three-phase system will be the same, and therefore they will be bouncing, or sliding, or vibrating in one spot with the same average speeds. The difference between them will be their inner heats—the heat of fusion and the heat of condensation.

If we add heat to the balanced system, first all the ice will melt without the temperature's going above 0° C. Then the water and vapor will rise in temperature until the boiling point, 100° C., is reached. It will stay there while the water boils away. When there is only vapor left in the container, the temperature will rise again to any degree we wish. So far we have not discovered a theoretical limit to heat; however, the molecules will begin to disintegrate, depending on the molecule, at definable temperatures.*

If we cool the hot water gas—always imagining it in the sealed container—at 100° C., it will start to become liquid, and the reverse process will take place, down through crystallization and the cooling of the solid to absolute zero. It is worth while to point out that every substance has its own boiling point and its own freezing, or melting, point and these are used as identifying physical properties. Water freezes at 0° C., platinum freezes, or melts, at 1775° C., oxygen at −218° C. But absolute zero is the same for all matter—which is an awkward way of putting it but brings the point home. Whatever the substance, whatever its particular freezing point, at −273° C. all its molecules stop vibrating.

There is a rapidly developing field of research into the properties of substances at this low temperature, called cryogenics. Startling changes occur that cannot be explained by the kinetic

---

* Such disintegrated particles have very strong electrical charges which repel each other violently but which enable the particles to be collected in powerful electromagnetic "bottles." A cloud of these enormously hot charged particles is called "plasma" and has properties which allow us to name it the fourth state of matter. Chemistry is not concerned with plasma to any great extent at this time.

molecular theory or by any other theory. Either our information is not sufficiently rich to develop a hypothesis for the dramatic change in the properties of matter near absolute zero, or perhaps we are trying to make sense out of what facts we do have by using the wrong point of view. Obviously the interior of molecules must contain the clue and, significantly, the madness that seizes them has no parallel. For instance, many nonconductors of electricity suddenly become superconductors near absolute zero. We have a wonderful explanation for electrical conductivity, but that explanation is totally useless when trying to make sense out of this dramatic switch from nonconducting to conducting properties.

However, we cannot expand on this here, and refer to it only to keep in mind the fact that even the kinetic molecular theory, whose importance to modern science cannot be overstated, has startling limits.

6

Vital Statistics on Molecules

Measuring the unimaginably small

Now that arguments for the existence of molecules have been presented, it is useful to list some of the factual data that have been collected about them in a variety of brilliant experiments which we cannot detail here but which are famous for their elegant simplicity.

The water molecule has a diameter of 4.5×10^{-8} centimeters, or $1.8/100,000,000$ of an inch. The mind cannot grasp this minute fraction. Some molecules are thousands of times larger, but even they are too small to be seen in any kind of optical microscope. To speak of such dimensions with greater ease the Ångstrom unit was established as one one hundred-millionth of a centimeter.

The speed of a water molecule at room temperature, whether as a vapor or a liquid, is about $1/3$ of a mile a second, which is a little faster than the speed of sound through air. The average distance any molecule moves in ordinary air between collisions is $1/160,000$ of an inch, another unimaginably small measure. The average gas molecule in ordinary air makes many many billions of collisions a second, the lighter ones moving faster making even a greater number.

Each molecule of water weighs 1.06×10^{-24} ounces, which means the fraction 1.06 over a denominator of 1 followed by 24 zeros.

A pint of water contains 1.89×10^{25} molecules, which means ten million times a million, times a million, times a million. One cubic centimeter, about a thimbleful, of ordinary air contains as many molecules as there are grains of sand in a pile one mile high and one mile square.

How is it possible to get these measurements?

They are the result of computations from data obtained with bulk matter. The first basic method worked out for obtaining any kind of numerical facts about molecules depends on a hypothesis proposed by the Italian physicist Amadeo Avogadro over a century ago.

Avogadro's principle

Imagine two totally empty containers, A and B, that have the same volume and that are kept at exactly the same temperature. Each has a tiny trapdoor just large enough to allow a molecule to pass in from the outside, and each trapdoor is manipulated by a creature so small that for him molecules are as large as tennis balls. The two tiny demons are also tireless. Each time one of these demons allows a molecule to enter his container, he signals to the other demon, who then also allows a molecule to enter his own container. In this manner the number of molecules in A always equals the number of molecules in B.

According to the kinetic molecular theory, since the temperatures are the same, the average kinetic energy of a molecule in A equals the average in B. Therefore the total kinetic energy of a hundred molecules in A will equal the total of a hundred molecules in B. The total force of the impacts in A will equal the total in B. The pressure in A will equal the pressure in B. Even if the molecules are of different size, small in A and large in B, the same situation exists, because we are talking about kinetic energies and not speed or size. The heavy ones in B will be moving more slowly than the light ones in A, and the average force of each impact in either container will be the same.

Avogadro's hypothesis, or principle, has been proved to be right countless times, but for a long time it was still called a hypothesis instead of a law, in memory of the fact that for fifty years no one believed Avogadro could be right. A precise statement of the law is: equal volumes of all gases, at the same temperature and pressure, contain equal numbers of molecules.

What is the usefulness of this law?

Suppose we weigh an empty container and, after filling it with a gas A, weigh it again, noting the pressure and the temperature. Then we empty the container and refill it with gas B at the same temperature and pressure, and weigh it again. Subtracting the empty weight of the container from the two other weighings, we get the weight of gas A and of gas B. The volumes were the same, the pressures and temperatures were the same. According to Avogadro's law, the number of molecules in gas A must have been equal to the number in gas B. If the weight of gas B is sixteen times the weight of gas A then, since the number of molecules were the same, each molecule of B must be sixteen times as heavy as each molecule of A.

This reasoning led to the determination of the comparative weights of basic particles. By assigning the number 1 to the lightest substance known, all others became multiples of 1. The standard originally chosen was hydrogen, which is the lightest of all substances. However, the method we have outlined can be used only with gases. The comparative weight of substances which cannot be easily handled as gas must be determined in other ways.

The actual weight of a basic particle could not be determined until the X-ray microscope was invented. In this instrument, X rays are deflected by the dense parts of the particles, frozen into a crystal, and fall on photographic plates as a pattern of light dots. A great deal of technical experience is required to translate those two-dimensional photographs into three-dimensional models of the crystal and then to calculate the distances between the centers of the dense parts in the crystal. Such data give a measure of the diameters of the basic particles. Having measured the crystal and calculated its volume, and knowing from the X ray the volume of a single particle, we can calculate how many particles there are in the crystal. The crystal is weighed and thus we obtain the total weight of a known number of particles. Simple division enables us to calculate the weight of a single particle.

For the chemist, comparative weights are far more important than actual weights of particles, because he always works with bulk matter.

In the laboratory, experiments with gases are not conducted under constant temperature and pressure, or ideal conditions, but the combined gas laws allow accurate corrections to be calculated. Under extreme conditions, the gas laws must be corrected because they are derived for an ideal gas, that is, for a gas whose molecules are mere mathematical points in space. Since gas molecules are real bits of matter, they have a volume of their own, and the space that the gas has available for its movement in a container is less than the total volume of the container by that amount of space which the molecules actually occupy. In other words, to use the gas laws with accuracy we have to subtract the volume of liquid a gas would form, if it were condensed, from the space available in the container. Also, a correction must be made to the ideal gas law which relates to the change in temperature that automatically cools a gas when it expands and heats it when it is compressed. And finally the attraction between gas molecules when they are very close must be accounted for in dealing accurately with real gases.

Many laws derived in the nineteenth century were for ideal systems. That is, reality is a complicated business, but it is possible to discern single aspects of reality and to extrapolate measurements of them into ideal conditions, in order to discover the fundamental principles underlying the surface confusion of things. Thus, the ideal gas which can be imagined in our model allows us to explain absolute zero as the temperature when all motion ceases. Real gases condense and freeze long before that temperature is reached, and we could not arrive at the concept of absolute zero purely on the basis of careful measurements of the relationship between temperature and volume.

One of the everyday uses to which the gas laws have been put is weather prediction. The whole theory of weather formation depends upon the kinetic molecular theory, according to which vast systems of air masses are created that have contrasting temperature and resultant pressure conditions in them. A myriad number of dynamic relationships exists in every air mass—the term "mass" here is not used in the strict sense—and even a slight change due to a mountain's poking into the mass or a movement over a forest or a small lake is enough to start chain reactions. There is a careening about in the skies of immeasurable, multi-layered, billowing seas of gases of all sorts and origins. Currents whip around the globe or sink back in vast flowing rivers or spiral upwards in violent storms. The kinetic molecular theory does not chart the convolutions in the air, but meteorologists apply the gas laws to measurements of temperature, pressure, wind, and humidity, taken at regular intervals, so that the changes in these can be plotted as evolving trends.

But now we leave the kinetic molecular theory and move inside the particles that we have been discussing as though they were undifferentiated, solid spheres.

PART II

The Atom in Close-up

•••

7

Atoms as Solid Particles

The precision of theories

NOTHING in the kinetic molecular theory explains chemical change. Nothing in that particular view of matter hints at the manner in which these basic particles, in constant motion, combine with one another in the countless chemical reactions that we know. Since they move about on their own, nature often allows one kind of particle to come in contact with other kinds, and thus provides them with an opportunity to react. Sometimes they do, sometimes they don't. In the laboratory it is easy enough to pour one sort of liquid into another or on a solid, or to strike two solids together or to let two or ten gases mix in a jar. Sometimes there is a reaction and sometimes not. The kinetic molecular theory cannot describe what happens when gasoline explodes, when a candle burns, when acid turns the red metal copper into a green liquid, when stew in the stomach is transformed into blood and bone, when a seed germinates underground, or how a child grows in the womb and keeps growing after he is born, or how anything dies. Everything we know about the world is known through molecular changes that occur in our nervous system as a result of stimulation of our senses, but the kinetic molecular theory does not even pretend to recognize the existence of chemical change, any more than the laws of gravity suggest how an egg is fertilized. The kinetic molecular theory concerns itself exclusively with the physical behavior of all basic particles, whatever their specific properties may be. No matter what the substance, its basic particles abide by the definitions of the theory. There is a freezing point and a boiling point, for instance, for every kind of molecule.

It sometimes happens in science that a man proposes a correct explanation for some phenomenon but it is ignored a long time by his colleagues. The cause is partly perversity of the human mind and partly the conservative nature of scientists, who, like most professional people, devote their energies to pursuits which do not need the illumination of new concepts. The fact is that most people are not creative in this sense, and scientists are no exception.

Nevertheless, two hundred years ago there was so much to discover with the scientific method that almost any sort of honest experimentation was bound to yield bits of new information. New substances and new chemical processes were being uncovered at an accelerating rate as part of the whole revolution in viewpoint that has created our modern world with its increasingly powerful techniques for harvesting more and more information about the universe, and for increasing our command over our environment.

The passion of the first part of the nineteenth century was for cataloguing sheer information, and many intuitive suggestions were ignored until the accumulation of data revealed their validity. Scientists felt that the future of science depended on a fanatical adherence to the scientific method—inductive reasoning from objective observations—and that mere speculations and hunches led the mind down a primrose path back into the intellectual morass of the Dark Ages. After all, for every great idea that might be neglected for a while because it lacked proof, erroneous ideas also died, but died a deserving death, from the same cause: lack of proof. Today scientists, more relaxed, have such a large bag of useful ideas without which their work would come to a standstill that they do not worry about their "scientific" proof.

It might be relevant at this point to define the essence of the quarrel in the nineteenth century between the sciences and all other philosophies. The critics of science said: nothing in this body of knowledge is permanent; all these theories and laws are constantly being changed to fit new facts; the mind is left restless and dissatisfied and longs for the eternal verities that all men can discover in their own hearts if properly instructed. The scientist replied: every word you say is true but what you don't say is that all philosophies and all religions promise knowledge of nature, but only science can fulfill that promise. It has done so just because it does not invent theories of what nature ought to be like and so it does not distort facts to fit the theory; science reasons from the facts it has available, and when more facts come in it adapts its theories to suit the new facts.

In the twentieth century the ground for this quarrel between science and the humanities has more or less vanished. The power of scientific knowledge is now so vast that scientists are being forced to face the problem of how to define their moral position on the use of such power. Scientific knowledge belongs to all men, and thus all men, including scientists, must accept a new kind of responsibility no one ever dreamed of until now: the responsibility of actually creating a new kind of civilization.

The atomic theory

But let us return to the beginnings of modern chemistry. All through the eighteenth century the vague idea of basic particles hovered in a few probing minds, mostly because it was a tantalizing intellectual nugget rather than because of any increasing evidence for their existence. The actual mechanics of chemical action, however, remained utterly baffling and seemed quite disconnected from the idea of particles. Finally the English chemist John Dalton in 1803 proposed an atomic hypothesis founded purely on the weight relationships that had been discovered in chemical reactions. But he could not explain with this theory a great deal of the known phenomena of chemistry, and for this reason it was not accepted generally until fifty years later.

The assumptions in Dalton's atomic theory have been revised radically more than once, and in fact today it is impossible to contain all our knowledge of the atom in a few simple sentences. We no longer need a theory concerning the existence of atoms, though we are in quite desperate need of a totally new concept of the atom's structure. But it is well to introduce atoms with Dalton's theory, updated, because it does formulate the essential argument for their existence.

NOTA BENE: Henceforth, "atom" and "molecule" have separate, precise meanings; we no longer use the word "molecule" as a generic term for all basic particles.

We begin with the statement that all matter in the universe exists in the form either of an element or of a compound which is constituted of elements in chemical union. Elements are composed of atoms, and compounds are composed of atoms of different elements bound into molecules.

Atoms are the smallest independent particles of an element that can exist and still retain all the properties of that element. Atoms of any one element are chemically all the same, wherever they are found or manufactured. Atoms of lead obtained from ore laid down millions of years ago are chemically identical with atoms of lead on other planets, or with atoms of lead produced in nuclear reactors, or with atoms of lead hammered into lead pipe. Helium in outer space is identical with helium pumped out of Texas fields. Atoms of different elements are different in unmistakable ways. An atom of lead is not much like an atom of helium; an atom of sulfur is not like an atom of iron.

Atoms combine chemically with one another under certain conditions to form molecules of compounds. A molecule is the smallest particle of a compound that can exist independently and still show all the properties of that compound. According to the laws of conservation, the sum total of all matter and of all energy associated with the atoms which combine with one another is equal to the sum total of matter and energy associated with the molecules which result from that combination. All the molecules of a certain compound produced anywhere are identical because the atoms always combine in the same proportions. Water molecules anywhere in the universe consist of 1 atom of oxygen (O) and 2 atoms of hydrogen (H). We have already stated that all atoms of hydrogen are chemically the same and so are all atoms of oxygen chemically the same. Therefore the same combination of these will always produce the same compound.

The same atoms can combine in different proportions to form different compounds. Hydrogen peroxide molecules contain 2 atoms of hydrogen and 2 atoms of oxygen, and this molecule, written in the symbolic formula as H_2O_2, is quite different from the molecule of water, H_2O. Carbon monoxide, CO, and carbon dioxide, CO_2, are altogether different; the first is fatal if we breathe it, and the other is not.

The ability of an element to combine in different proportions with the same partner was one of the more confusing aspects of chemical reactions, and Dalton could not imagine a linking mechanism between atoms that allowed for the same carbon atom to tie up two oxygen atoms under one kind of circumstance and only one oxygen atom under different circumstances. Nor could he imagine how these atoms, simply by linking into molecules, produced totally different kinds of substances. The elements hydrogen and oxygen and the compound water are three utterly different kinds of substance.

Elements, compounds, and mixtures

Over the years, researchers have found and studied 92 elements occurring naturally on earth, in the earth's crust, its waters and gases, some of which are radioactive, and have created in nuclear reactors a dozen more which are radioactive, some existing only for fractions of a second before decomposing. The whole universe is made up of these particular 103 elements. Some of them are familiar as the common metals iron, copper, silver, gold, lead, tungsten, chromium, platinum, aluminum, tin, and so on. Oxygen, hydrogen, nitrogen, helium, neon are elements which exist as gases at ordinary temperatures but which can be cooled into liquids and solids, just as all metals can be melted into liquid and vaporized. No liquid elements exist at normal temperatures, except mercury and bromine. Sulfur is one of the few elements which occur in pure deposits in the earth, and the element carbon in the shape of diamonds is another. But most of the elements found in the earth's crust and waters are combined with one another into compounds, such as ordinary table salt, limestone or chalk, sand, and the numberless minerals that geologists study. From these compounds the pure elements can be obtained by a variety of chemical processes.

If each element combined only in a single way with each of the other elements, we would have over eight thousand compounds. But each element combines in various proportions with many other elements and, furthermore, there are compounds with three, four, five, and more elements combined, and some molecules have thousands of atoms of three or four kinds in them. Altogether the known compounds found in nature and manufactured in laboratories are estimated at around a million. Theoretically the number possible is infinitely large.

The ancient Greeks had four or five basic elements, the alchemists had five or seven, and then for a long time the idea of elements was both philosophical and realistic, but always it resisted any intrusion of the atomic theory. The idea that elements were solid matter rather than abstractions spread steadily after 1800, but the idea of compounds remained confusing until the evidence for them was overwhelmingly clear. Then Dalton's theory was recognized, and with it everything fell into place. No other explanation could fit the facts. The age-old tussle between "elements" and "atoms" ended with both concepts proving to be right in a combined concept.

Practically everything in nature is composed of compounds, or a hodgepodge mixture of compounds. Very few elements can exist for long under normal conditions in an atomic state. We have mentioned sulfur, and carbon in the form of diamonds; gold is another one of the few, and so is helium.* If we analyze or break up the compounds that constitute the crust and waters of our earth, if we wrench the atoms out of their molecular combinations, we find by estimation that 50 per cent by weight is oxygen. By this we mean that if the crust and waters of the earth could be weighed and the molecules of the hundreds of different compounds that form the familiar rocks and dirt and sand were broken down into the atoms of the ninety-two elements, and all of the freed oxygen atoms were collected, they would constitute about one half of the total weight.

In other words, the atoms of all but a few elements are found in nature chemically combined with one another into molecules, and a large number of these molecules also contain oxygen. The atoms of the element silicon, Si, combined with other atoms are estimated to be 25 per cent of the earth's crust by weight. Ordinary sand is silicon dioxide, SiO_2. The element aluminum, Al, combined in a variety of ways with other elements, constitutes about 8 per cent of the crust. Most clays contain aluminum. Iron, Fe, is 5 per cent, calcium, Ca, is 3 per cent, magnesium, Mg, sodium, Na, and potassium, K, make up 2 per cent each. All the rest of the ninety-two elements are estimated to be only 1 per cent. Included in this 1 per cent is carbon, C. Every molecule of every living cell, vegetable or animal, contains carbon atoms, yet the weight of all living organisms on earth is a tiny fraction of 1 per cent of the weight of the crust.

* Some other gases exist as molecules which are not compounds because they contain only atoms of the same element bound together. For example, oxygen in the atmosphere consists mainly of molecules containing two oxygen atoms, and there are also traces of another form called ozone, whose molecule is made up of three oxygen atoms.

If we look at the whole universe, however, the most common element in it is hydrogen, and the next is helium.

In chemistry, a molecule cannot be spoken of as a mixture of atoms. A molecule consists of atoms combined in a fixed proportion. That is, each kind of molecule always contains identical numbers of the same atoms. Water is always two atoms of hydrogen, H, and one of oxygen, O. But a mixture consists of any number of elements or compounds in any kind of ratio. Air in a room with a faulty stove will contain far more carbon dioxide and carbon monoxide than it does normally. Any room with people in it will contain less oxygen and more water vapor than it will without people. Air is a mixture of gases, whose proportions can be varied endlessly. However, the over-all atmosphere of the earth is remarkably consistent in the proportion of its gases because of the winds that blow constantly. Sea water is a mixture of water and dissolved solids, the percentages of which vary from place to place, being less salty near the mouths of rivers. Tap water is a variable mixture of water and dissolved minerals. Granite is visibly a variable mixture of many kinds of stone stuff. The silver used in tableware and money is a mixture of the elements silver and copper, and others. Steel is a mixture of iron, carbon, manganese, and other metallic elements in varying proportions.

Mixtures in which the ingredients are uniformly mixed, like ink and water, or tea and sugar, or air, are homogeneous and are called solutions. In a solution, gaseous, liquid, or solid, any sample will contain uniformly distributed atoms and molecules, but they are not even slightly in chemical bonds with one another. Any mixture in which the ingredients are visibly contrasted, like sands on a beach or oil and water shaken up, is not homogeneous and is not a solution. Smoke and air are not a solution but are a mixture.

The study of mixtures and solutions has created a whole field of knowledge extremely important in every industry, for the obvious reason that almost never are elements handled pure.

To repeat, in a compound the atoms are chemically bonded, in mixtures and solutions atoms or molecules are simply thrown all together in any proportions and are not chemically bonded.

To illustrate the problem of imagining what happens when atoms combine, consider the following facts. The element sodium (Na) is a metal with a silvery color; it melts at 97.5° C. and is lighter than water; it reacts violently with water to produce hydrogen gas; and it is poisonous to the human system. Every atom of sodium has these properties. The atoms of the element chlorine (Cl) are also poisonous, fatal when breathed. Chlorine is a greenish gas which condenses at —34.6° C. and solidifies at —101.6° C. The gas consists of molecules containing two atoms of chlorine, and is soluble in water, with which it forms the powerful hydrochloric acid.

If a piece of sodium is put in a container of chlorine gas there is a violent reaction with flame and heat. The product of the reaction is a white granular solid that has a melting point of 804° C., and it is wholly different from either sodium or chlorine: it is nothing like any metal; it does not react with water, though it dissolves in it. And it is absolutely necessary for life: it is ordinary table salt, sodium chloride, NaCl, consisting of one atom of chlorine to every atom of sodium. When these atoms are separated from their chemical bonds and set free, they are again poisonous sodium metal and poisonous chlorine gas.

Obviously the explanation must be found within the atoms themselves, and, in fact, during the last half-century a totally unsuspected structure has been slowly uncovered, a structure even less predictable from the surface appearance of things than is molecular movement.

The information about the atom's interior was collected in three quite different areas of research, and the researchers were not collaborating in any organized way. In fact they were asking questions and designing experiments that only vaguely related to the idea that the atom might have a complicated structure. Gradually their findings began to overlap; one brilliant mind after another began to recognize similarities and analogies and saw the relationships with the kind of boldness that brushes aside logical barriers to the conceptual grasp of the essential truth.

8

The Theory of the Solid Atom
Becomes Untenable

The original idea of valence

THE first area of research that yielded facts and ideas about the atom was, of course, chemistry itself. Chemists compiled a list of comparative atomic weights, with hydrogen, the lightest element, being given a weight of 1. Not one this or one that, but just 1. Oxygen, being 16 times as heavy, had an atomic weight of 16. But for some time after this notion became absolutely necessary to chemists in the nineteenth century, no one could imagine how one atom could be actually weighed. That came only after World War I.

The much more difficult task of arriving at some idea concerning the combining power of an atom also belonged to chemistry. The resulting theory of valence, which originated about a century ago, was destined to undergo several radical changes. Broadly speaking, valence is the combining power of an atom. At one time valence was imagined as a hook. One, two, three, as many as four hooks on an atom would constitute a valence of 1, 2, 3, or 4. An atom with a valence of 3 could seize one other atom with a valence of 3. The 3 hooks on each would clasp the 3 on the other neatly. An atom with a valence of 3 could hook to itself three atoms if they each had a valence of only 1. Quite complicated combinations could be worked out among three atoms, each with a different valence. The point was that all the hooks had to be hooked up in a stable molecule, and though the hooking pattern of many compounds could not be worked out, the concept of valence and the use of valence numbers became the very heart of chemistry. It established absolutely the fact that no parts of atoms were lost during chemical change—that always whole atoms combined into molecules.

Chemists emphasized at international conferences their solid conviction of the existence of atoms, and this insistence was useful to other scientists. But chemists could not get inside the atom for the simple reason that the study of chemical behavior is concerned with bulk matter rather than with single atoms. The other two sources of information about atoms were electricity and radioactivity.

Digression into physics

The ancients knew quite a bit about static electricity, which is the kind that discharges from your fingertip when you touch a metal doorknob or radiator after shuffling across the rug. No one knew more about electricity than the ancients did, until the eighteenth century.

Elementary courses today show how a glass rod rubbed with silk, or an amber rod rubbed with wool, will pick up bits of material the way a magnet picks up bits of iron. A series of simple and entertaining experiments with a small pith ball on a thread shows that apparently there are two kinds of electricity. Matter that contains kind A is repelled by any matter that also contains kind A. Anything that contains kind B will attract anything containing kind A. That is, oppositely charged bits of matter will attract each other, while bits of matter charged in the same way will repel each other.

Magnetism seems to follow an identical pattern of behavior: the same poles of two magnets repel each another, but opposite poles attract each other.

Benjamin Franklin suggested that there was only one kind of electricity and that neutral objects have a kind of balanced condition. When some of the electrical stuff is drained away it leaves the object in a condition that will attract to itself another object that has an excess of electrical stuff in it. Two objects both drained will repel each other. Two objects both overfilled will also repel each other.

The problems involved here, and the problem of trying to define this electrical stuff, finally resulted in a convention that electricity itself is negative. Therefore an object with an excess of electricity was said to be negatively charged and was given a minus sign, while an object with a lack of electricity was said to be positively charged and was given a plus sign. All pluses repel all other pluses; all minuses repel all minuses. All pluses attract all minuses. Neutrality is zero.

This convention had been worked out and was standard in everyone's thinking by the early nineteenth century, together with the realization that electricity could flow along a conductor or collect in condensers and that the behavior of magnets when they were placed near conductors, or near flows of electricity, was somehow related to the behavior of electricity.

The electron discovered

Among the many kinds of electrical and magnetic phenomena that fascinated physicists, who invented many wonderful devices to illustrate the strange properties of electricity, was the spark that leaped through the air between two highly charged—oppositely charged—electrodes.

This discharge of electricity through a gas produced certain observable reactions in the gas itself; for instance, an electric discharge produces different colors in different gases. Even more remarkable, differences were obtained by sending discharges through a partial vacuum, which was created by pumping air out of a glass tube containing negative and positive electrodes sealed into its opposite ends. The tube could be filled with pure gases at different pressures. Sir William Crookes is credited with refinements in the making of this apparatus, called the Crookes tube, and before the end of the nineteenth century Joseph John Thomson, later Sir Joseph, refined the experiments made with the tube.

When a high-voltage electrical current is applied to the ends of a Crookes tube filled with any kind of gas, the electric charge which, by convention, is called negative builds up on the tube's negative electrode—the cathode. When the charge is sufficiently high it discharges in a sudden burst across the tube to the anode, which is the positive electrode. As electricity is fed to the tube, the charges build up on the cathode and bolt across to the anode. If the gas is pumped out of the tube, instead of an intermittent series of spark discharges, a steady greenish glow appears on the tube wall opposite the cathode. Furthermore, the position of the anode has no effect on the discharge; even when the anode is sealed into the side of the tube, the glass wall opposite the cathode still fluoresces with a green color, indicating that something emanating from the cathode hits the glass and causes the glow. This "radiation" was named cathode rays, and further investigations yielded the following conclusions concerning their nature.

An object placed in the path of the rays cast a shadow on the fluorescing glass wall of the tube—a shadow that was a faithful silhouette of the object. This could mean only that cathode rays traveled in straight lines from the cathode.

Next, a little paddle wheel, placed on rails that ran the length of the tube, was propelled away from the cathode when the radiation hit the paddles. This proved that the rays must consist of solid particles, or the paddle wheel would not move.

Finally, a magnet held close to the tube bent the cathode rays from their straight path, proving that they consist of electrically charged particles. Hence, the rays are also deflected by electric charges, and the direction of the deflection shows that the charge carried by the cathode-ray particles must be negative and not positive.

It was legitimate to conclude that cathode rays consist of particles of matter that have mass and volume and a negative charge. These particles were named electrons, and the amount of electricity that they carry, the magnitude of their electrical charge, was considered to be the smallest possible quantity of electricity that could exist by itself.

Since the rays stopped when the current was shut off, these particles had to come from the material of the cathode. Thus electrons must be part and parcel of matter itself and related to matter in such a way that normally they cannot be detected in neutral objects.

Every atom, and therefore every molecule, which consists of atoms, must contain electrons that can be pried out and sent speeding through space. The remaining bulk of the atom, robbed of some of its electrons, must be positively charged. Such positively charged atoms or molecules would be attracted to the cathode or to any negatively charged object.

Following this reasoning, the Crookes tube was searched for positively charged particles and they were found—a thin trickle of them moving away from the anode toward the cathode. Various experiments left no doubt that if the tube contained a little bit of gas, the streaming electrons of the cathode "ray" hit the gas molecules and knocked out of them a few electrons. The gas molecules were now left positively charged and moved to the cathode.

Even after these phenomena became familiar to all scientists, no one could propose a theory to explain the electrical nature of atoms.

The disintegrating atom

Clues to the complexity of the atom came from another field altogether, as a result of the discovery of radioactivity in 1895 by the French physicist Antoine Henri Becquerel. By the turn of the century it had been established that uranium and several other newly discovered heavy elements spontaneously emitted three kinds of radiations, which were named alpha, beta, and gamma rays.

Alpha rays consist of a stream of positively charged particles with a mass four times that of hydrogen. Beta rays consist of negatively charged particles identical to the electrons of the cathode ray. Gamma rays do not consist of particles at all but are electromagnetic waves like light, only of much higher frequency, which is the same as saying of much shorter wavelength.

Plainly, both negative and positive particles must exist in atoms and can be found there without benefit of electrical currents. This was absolute proof of the electrical nature of atoms.

More and more experiments with electromagnetic apparatus and with radioactive elements accumulated evidence that the atom was not a solid little nut at all, but consisted of even smaller particles electrically charged and held together at least partly by their opposite charges.

Several models of the atom were proposed to illustrate what was known, but let us consider only the one that has become most useful for modern scientists. However, right from scratch we must keep clearly before us the fact that although we can talk about a model for the atom, we cannot in any sense imagine the real atom.

We have already quoted some undigestible figures concerning molecules. The diameter of a helium atom is about 2 one-hundred-millionths of a centimeter. Alpha particles, which are actually helium atoms stripped of their electrons, are emitted by radioactive elements at speeds from 5 to 7 per cent of the velocity of light, which travels at about 186,000 miles per second in a vacuum. Electrons are only a fraction of the size of the atom, and those ejected from atoms in the form of beta rays may travel with velocities as high as 95 per cent of the speed of light. It is quite plain that even in this first glimpse inside the atom we see forces at work beside which the familiar forces of nature are incomparably tame.

A famous experiment on the scattering of alpha particles has to be described for several reasons. It showed that methods could be devised for handling single atoms; it supplied absolute proof that certain models of the atom were wrong; it focused all atomic research in a direction that eventually produced the first cyclotron or "atom smasher." Finally, it is an example of the brilliance that is necessary in science to keep abstract concepts grounded in solid matter.

The British physicist Ernest Rutherford was already famous for his researches in radioactivity when he set up the experiment in 1911. He bombarded a target of very thin gold leaf with alpha particles emitted by a radioactive element. We have mentioned that these alpha particles are four times as heavy as hydrogen atoms and that they are in reality positively charged residues of helium atoms, and that they travel at a speed of roughly 10,000 miles a second. Most of these particles went right through the gold leaf to hit the phosphorescent screen behind it in direct line with their source, just as though the gold leaf had not been there —as though the gold atoms of the leaf were empty space. Some of the particles, however, were deflected to the side and hit the screen at different distances from the bull's-eye. And a few of the particles were knocked clean backward without ever going through the gold leaf.

Rutherford's explanation—and we have not indicated any of the prevailing ideas and information which helped him interpret these facts—was the following. Atoms of gold were so close to-gether in the solid gold foil that not even light rays could get through between them. Since positively charged helium atoms did penetrate the thin foil, they must have gone right through the atoms, not between them, and yet the gold leaf was not damaged. There must be empty space inside the atom, enough empty space to allow an alpha particle free passage. Some of the alpha parti-cles, however, were knocked aside. Hence, there must be within the atom's total space a very solid part. After all, the gold atom weighs about fifty times as much as helium. And this solid part must have a positive charge on it because a few of the positively charged bullets were flung right back, more violently repelled than could be accounted for by a direct hit on the solid part.

The model Rutherford imagined consists of a center core, a positively charged nucleus that contains most of the mass of the atom. Since the atom as a whole is neutral, electrons are grouped around the nucleus in such a way that they balance the positive charges. These electrons are at a relatively considerable distance from the nucleus, and it is these electrons that create the surface of the atom.

This model, adapted by the great Danish physicist Niels Bohr to account for the electron structure of the atom, is still the skeleton which we clothe with what contemporary knowledge, vastly superior to Rutherford's, it can take on. Apart from guidance in teaching, the Rutherford-Bohr model is so inadequate that its use is justified only because so far nothing better has been proposed. However, it does serve excellently to introduce the mind to the energies within the atom, and with some modifications it will always be a fine model for chemical reactions.

9

A Preliminary Model of the Atom

The simplest atom of all

T H E simplest diagram of the atom that we can draw from our information in the previous chapter is that of a circle with a dot at the center. The dot represents a very small and positively charged nucleus which contains practically all the mass of the atom. The circle at a distance represents the atom's skin, as it were, which consists of electrons. Since the mass of an electron is $\frac{1}{1850}$ of that of a hydrogen atom and hence is relatively negligible, one may say at this point that the outer surface of an atom consists of a skin of electrons and that it is the electrons which come into contact when two atoms touch, when any object touches another.

Atoms are electrically neutral, and therefore the picture of the simplest atom that we can imagine is one with a nucleus that has a single positive charge balanced by a single negative electron at some distance from it. This is, in reality, the hydrogen atom, the smallest and lightest atom of all, to which was assigned an atomic mass of 1 over a century ago. Two of these hydrogen atoms linked in chemical bonds constitute a molecule of hydrogen, H_2, which is the gas as we know it at ordinary temperatures.

The atoms of all other elements are heavier than hydrogen and therefore have larger nuclei with more than one positive charge and thus with more than one electron at a distance from it.

The diagram of the dot with the circle around it is, as we have cautioned, quite misleading in many ways.

In the first place, if the positive charge on the nucleus and the negative charge of the electron attract each other, why doesn't the electron fall into the nucleus? The answer is that the electron is moving at speeds close to 100,000 miles a second, which would carry it four times around the earth's equator in one second, and this speed gives it a centrifugal force that exactly balances the pull between the opposite charges. This explanation has to be modified considerably to take care of most recent discoveries, but it will serve us very well for a while as a kind of analogy.

(Centrifugal force should be defined here. If you swing a stone on the end of a string around your head it will seem to weigh more, the faster you swing it. If you swing it fast enough the string will break and the stone will fly off at a tangent to the circle through space. The earth's gravitational pull will make the stone curve down toward it, but that is an influence on the stone which has nothing to do with its breaking the string. The string can break when the stone is at the top of the swing. Why did the stone become so heavy simply because it was moving faster? Its mass did not increase, but its inertia to travel in a straight line did increase. This is the principle of the slingshot, or of discus throwing. This is why water stays in a bucket when it is whirled around fast enough. The force that keeps the water pressed against the bottom of the bucket when the bucket is upside down as it swings overhead—pressed there despite the pull of gravity— this force is supplied by the energy of centrifugal speed. Now imagine that instead of a stone on a string we have a rocket moving under its own power on the string. It will want to travel a straight line through space. The string holds it. It will therefore sooner or later arrange itself in such a way that it will travel at the end of a string in a circle.)

The electron is moving under its own steam too, just as mysteriously as the incomparably slower but ceaselessly moving gas particles, and will travel in a straight line if it can. But it is held by the positive-negative forces of attraction between electron and nucleus. This attraction keeps the electron from flying off into space, while the centrifugal force of the electron provided by its own speed keeps it from crashing into the nucleus.

In all our discussions this balancing of opposed forces must be kept in mind; otherwise the image of the atom tends to simplify in the mind and to degenerate into a couple of marbles swinging about each other on the ends of a wire.

To return to our dot-and-circle diagram of the atom: the circle indicates the path the electron takes in its race around the nucleus. But atoms are not flat plates but three-dimensional rounded objects something like a sphere. Thus the electron weaves a whole pattern of orbits around the nucleus. Our diagram should therefore indicate with an arrow that the circle is an orbit going in one direction, and there should be a great many of these orbits drawn. We let a single circle represent them all.

Both electron and nucleus are spinning on their own axes, and this ought to be indicated in the diagram too.

The various orbits of an electron are not all circular but sometimes flatten out into ellipses, twist into dumbbell shapes, and even elongate into long swings out and in. This again is difficult to diagram; however, we can calculate with absolute certainty, from analysis of crystal structure in which the atoms are rigidly locked insofar as their nuclear distances are concerned, the average distance the electron maintains from the nucleus. A solid does not bulge and shrink, billow, or cave in because of the changing orbits of electrons around the nuclei. The scientific way of saying this is that at any instant the chance of finding an electron is greatest at a certain distance from the nucleus, a distance that can be measured by using X rays in a way to cast a complicated shadow of the nuclei on photographic film. The electron at times might be much farther away, but the chances are mathematically such that over a period of time it will be found most often at the stated distance.

The distinction between matter and energy on this minute scale has almost vanished, in a manner of speaking, but it still exists, and in order to explain our mathematical knowledge of the electron's behavior we must "think" of the electron not just as a speck of matter but also as a wave of energy. Not only is it impossible to diagram this concept, but it cannot really be imagined. Certain properties of waves preclude the image of a finite particle, while certain properties of matter preclude the image of waves. We must, as with light, employ sometimes one and sometimes the other concept.

Any electron in any atom can be excited by collision with an outside source small enough to be on the same scale. Radiations, such as X rays, other fast-traveling electrons, or another and more energetic atom, can whack an electron into traveling faster, thereby giving it greater energy to move farther away from the nucleus. But it moves farther away in leaps rather than in a smooth spiral. That is, the electron can occupy any one of several well-defined orbits, or tracks, around the nucleus, depending on how much energy it has, but it cannot revolve between those proscribed orbits. It was Bohr who reasoned that the electron simply leaps from one groove, or orbit, into the next if it is excited to do so. It seems to vanish from one level of energy and reappear at a higher energy level farther from the nucleus if something has given it the extra juice, or in a level nearer to the nucleus if it loses some juice. If left alone an excited electron will gradually fall back, erratically, perhaps, level by level into its normal channel, the so-called ground level, emitting as it leaps down the energy that it had absorbed when it became excited. How can any of this be diagrammed? Even verbalization distorts the known mathematical relationships.

Finally, if we enlarge our model until the nucleus is the size of a baseball, then the electron in the hydrogen atom will be circling the nucleus 100 yards away. It is impossible to reduce this scale to a dot and small circle.

We have only begun to detail the machinery inside the atom, and already our model is hopelessly inadequate to carry it. And yet it is inestimably useful for further discussion.

First we enlarged the surface of bulk matter until we saw the leaping, bouncing atoms and molecules of a gas slowing down to sliding and bumping in liquids and then becoming fixed in solids but still vibrating, though less and less, until at $-273°$ C. all motion ceased. And whatever the atom or molecule may be, this picture is the same. Next we enlarged the atom, any atom, and found practically empty space in it, with almost the whole mass of the atom concentrated into a tiny, seemingly impenetrable heart or nucleus and the shell-like outer surface of the atom of almost no mass made up of the ghostly weaving of electrons: a cloud. We still have no idea of how two bounding atoms with their clouds of electrons link together into a molecule. But before we proceed we must complete the list of subatomic parts, insofar as they are known today.

Exploring the nucleus

After Rutherford's classic experiment on the scattering of alpha particles, further research in which neutral atoms were bombarded by alpha and beta particles shot from radioactive elements eventually led to the construction in the 1930s of the first "atom smashers." From these machines and modern developments of them, and from atomic reactors, we have extracted proof that there are more than thirty bits of basic matter, and energy, associated with the nucleus. The most familiar of these elementary particles are the proton, neutron, neutrino, meson, and positron. For each of these particles, and some have several versions, there is an anti-particle; a collision between a particle and its anti-particle destroys both, creating energy.

Nobody can guess how all these pieces are assembled into the nucleus, or if all exist in the stable nucleus not under bombardment. At first it was thought that perhaps mesons act as a kind of cosmic glue, welding together the positively charged protons, which otherwise would repel each other with enormous force. But the idea of a glue does not advance us at all—obviously the positive charges do stick together with a force greater than the normal repulsion between them—and to call the process a gluing is merely to give it a homey name. Furthermore it might convey the wrong connotation because instead of a glue, which would have to be a fragment of matter or energy, it is possible that the configuration of the fields of force around the particles permits or brings about a locking together.

In any case, only the proton and the neutron and the electron have an enduring existence as independent entities on the face of this earth, and the other pieces have so far manifested their presence only fleetingly, to expire in mysterious ways. The specific properties that have been learned concerning these three enduring parts of an atom can be described in rigorous terms as follows.

The proton, which was discovered by Rutherford in 1919, has a mass of approximately 1, relative to oxygen 16. In chemistry and in much of physics the exact mass of these particles is not used. The proton carries a single positive charge of electricity, equal and opposite to the charge on the electron. As a result, it can be pushed and pulled about in electrical and magnetic fields. Its speed is dependent on the local conditions in which it finds itself, and normally it approximates the speeds of ordinary gas atoms. Protons are the nuclei of hydrogen atoms.

The neutron has a mass almost but not quite the same as a proton's, but in the calculations of chemists and most physicists its mass is considered to be 1. The neutron has no electrical charge, and so it was not discovered until 1932, many years after the proton was quite well known. Being neutral, the neutron cannot be handled in electrical and magnetic fields. Under normal conditions it behaves not at all like any ordinary gas, because, having no charge, it slides right through atoms.

The electron has a mass approximately $\frac{1}{1842}$ that of the proton. It has a single negative charge and is thus affected by any electrical and magnetic field in its vicinity. The speed of an electron depends upon its environment, but it can approach the speed of light and is always in that range of velocity.

For the chemist only these three elementary particles—protons, neutrons, and electrons—are important, as they are the stable constituents of the atom. With these we can explain nearly all the events in a chemical reaction, all physical properties, and all the grand cycles of nature that interest us at this time. The information on things which we feel we cannot explain is growing steadily, and sooner or later the inexplicable, or paradoxical, or utterly unrelated data that scientists are collecting about nuclear particles will yield a pattern in the mind of some genius, or something in that pile of data will nag at a lot of people to create a mosaic of that pattern, and a formula will be proposed by someone to embrace all the data in a way that arranges the mishmash of several formulas into a beautiful theory. Quite possibly the brief life of splinters from a shattered atom might be the very clue to a deeper understanding of chemical reactions. But for the time being we don't know the significance of the other particles and must ignore all but the electron, the neutron, and the proton. Those who would like to speculate on a new model for the atom should be warned that even the Rutherford-Bohr model is structured on the most difficult mathematical formulations the mind knows.

A general picture

Let us imagine we are being taught the anatomy of the fly. We have watched and killed countless numbers of flies since childhood. Now very thin slices of the fly, stained with dyes, are slid into a microscope, and when we have seen all the slides in order we can draw a picture of the fly's anatomy. Of course we'll never be able to visualize that anatomy actually working away to fly the fly about, to make it eat, copulate, lay eggs, and die. Still, we will know a great deal about the working parts that no amount of contemplation from the outside could reveal. The same situation exists in every science, but in physics and chemistry the analogy is stretched to the limit.

If we contemplate the general picture we have developed from the first paragraph of this chapter, we now see protons and neutrons densely packed into a spinning, positively charged nucleus while a cloud of almost weightless, negatively charged electrons swarms about the nucleus at a far distance from it. The energies in this system of spinning particles must be balanced in such a way that they keep the atom permanently stable over a wide range of conditions. All that flashing movement is fitted together into the rigidity of stone, the edges of swords, the flow of waters, the breeze—into the three-dimensional solidity of matter and energy as our senses know them, the solidity of our own flesh.

Yet we cannot slice up an atom. We have to observe multitudes and glean the nature of one by clever deductions from statistical measurements.

As we have already mentioned when discussing the kinetic molecular theory, the great difference between nineteenth- and twentieth-century views of nature lies in the conscious importance that we now give to the statistical realities of nature. One of the most important theories that uses statistics is Heisenberg's principle of indeterminacy. It states that, though we can measure accurately either the speed or the position of an electron, we cannot measure both at the same time with great precision. It follows that we cannot actually predict what any single electron will do, since we cannot get an accurate measure of both its present speed and position. But we can predict limits to its behavior and we can predict very accurately the average behavior of a large number of electrons. Since predictions concerning the future of an electron cannot be even theoretically better than a probable average of all its possible futures, the nineteenth-century clockwork universe, in which the position and speed of every particle could be theoretically charted, is no longer acceptable. The real atom is not a tiny, glistening, jeweled watch whose cogs mesh with inexorable precision into a ticking to the end of time.

It should be kept in mind that our information about the atom is compiled from the positions of needles quivering on instrument panels that we invented, from lines crisscrossing an exposed photographic emulsion that we invented, from sparks on a phosphorescent screen that we constructed, from threads of mist in a fog chamber that we invented, from bands of colored light on ground glass, and so on. And it is from the behavior of these recording instruments that we create possibilities that might explain the cause of their behavior.

10

The Anatomy of Individual Atoms

A difference in the number of parts

THE smallest particle of any element that still has all the properties of the element is its atom. There must be, therefore, ninety-two different kinds of atoms. Since each atom consists of protons, neutrons, and electrons, and a lot of other particles that barely exist outside the nucleus, we can say that the whole universe is constructed out of these subatomic pieces, put together in ninety-two different ways, which in turn are combined chemically into a theoretically infinite number of compounds. The dozen new elements created in nuclear reactors also consist of the same particles put together in a dozen more different ways. A proton from one of these is indistinguishable from a proton ripped out of the nucleus of a gold atom or of an oxygen or a uranium or any other kind of atom here on earth or in outer space. Similarly, all electrons wherever they are found are exactly alike, whether in a Crookes tube or an iron atom or as a beta ray emitted by a uranium atom. Any neutron, in or out of any atom, is like all other neutrons in every way. The only difference between electrons is their speed. The only difference between neutrons is their speed, and the only difference between protons is their speed.

The question immediately arises: are these basic particles put together in 103 totally different ways, or is there some kind of organizational similarity between a gold atom, for instance, and an iodine atom? Is there, in other words, a pattern for arranging the basic parts within a general structure? Is there an inherent relationship between each kind of atom and all the other 91 (or 102) kinds, apart from the fact that all are constructed of the same few pieces?

Early in the nineteenth century, long before anything was known of atomic structure, the idea tantalized some scientists, and especially the British chemist William Prout, that the lightest element of all, which seemed to be hydrogen, might be the universal stuff out of which all the other elements were constructed and that thus all the elements were intimately related. The idea was still another version of the Greek view that underlying all substances was a universal stuff, a quintessence. It proves how powerfully and consistently the Western mind has yearned for this simplification. In such a view, an atom of oxygen would be simply 16 atoms of hydrogen fused together, because oxygen weighs 16 times as much as hydrogen. But because so few elements had atomic weights that were exact multiples of hydrogen the idea of a universal element remained unproductive. For a while protons and electrons were hopeful candidates, then neutrons had to be added, and then thirty-odd more "basic" particles. But let us proceed to put together the atoms.

It is logical to begin with hydrogen (H), which consists of 1 proton and 1 electron and has no neutrons. If any proton and any electron from any source are brought together under ordinary conditions they will instantly link together into a hydrogen atom. An arrangement of their fields of force results which fuses their separate and altogether different properties in such a way that a completely new set of joint properties comes into being. These new properties are not just the sum of the old ones but belong exclusively to the coupled particles as a unit, and that unit is the hydrogen atom. Its mass is considered to be 1 in all chemical computations.

The next heavier element is helium (He), whose nucleus consists of 2 protons and 2 neutrons and thus has a mass of 4. Revolving around the nucleus to balance the 2 positive protons are 2 electrons. Lithium (Li) is the third element, with 3 protons and 4 neutrons in the nucleus and, of course, 3 orbiting electrons to balance the positive charges. The mass of the lithium atom is 7. Beryllium (Be), the fourth element, which has 4 protons, 5 neutrons, and 4 electrons, has a mass of 9.

It seems that the elements can be lined up in such a way that each one has exactly one proton more in its nucleus than the previous one, and also one more electron to balance it. The neutrons do not increase in any regular manner and do not require electrons to balance them since they are neutral. But they do add weight.

Actually, every atom of a particular element always has the same number of protons, but there may be a varying number of neutrons associated with the fixed number of protons in the nucleus. All oxygen atoms have 8 protons, since it is the eighth element in the line-up. But some oxygen atoms have 7 neutrons, some have 8 neutrons, and some 9 neutrons. All these atoms with 8 protons, no matter how many neutrons, are called isotopes of oxygen. All these isotopes exist in nature in a mixture that does not vary in proportion from one place to another and that gives an average mass for the three of 16. The isotope with a mass of 16, that is, with 8 neutrons, happens to be by far the most common of the three. In other words, if a sample of oxygen taken from the air is analyzed, a very small percentage of that oxygen will consist of atoms containing 7 neutrons, a very small percentage of the atoms will contain 9 neutrons, and the majority will contain 8 neutrons. Their masses will be 15, 17, and 16. The percentages in the mixture is amazingly consistent, no matter where the oxygen is found.

More will be said about isotopes later. Just now the point to remember is that the atoms of each isotope of an element are alike but the different isotopes of an element have different numbers of neutrons. The identifying feature of any element is the number of protons in the atom, which is the same in all isotopes of the element.

If this pattern of protons increasing one by one persists for all the 103 elements, then nitrogen (N), the seventh element in our line-up, should have 7 protons and 7 electrons. It does. Argon (A), the eighteenth element, should and does have 18 protons. Uranium (U), the heaviest of the 92 elements, turns out to have exactly 92 protons in its nucleus, together with 92 electrons in its cloud around the nucleus. It also contains 146 neutrons—that is, the most common isotope of uranium does—and so its mass is 92 + 146, or 238, compared to 1 for hydrogen. The 92 electrons have a negligible mass.

Thus we can conclude that, although the properties of each element are recognizably and uniquely its own, and quite different from the properties of other elements, so that oxygen is nothing like iron and sodium is nothing like sulfur, there is nevertheless only a mathematical difference between the structures of the atoms.

This was certainly not foreseen when work on the inside of the atom began just before World War I, and it is rather unbelievable that the elements really are created with this extraordinarily simple relationship to one another.

A little clarification of terminology is useful at this point. The atomic number of an element is the number of its protons. The atomic mass of any atom is the sum of the masses of its neutrons and protons; except in nuclear research, both protons and neutrons are considered to have a mass of 1. The mass number of an isotope is thus the number of its protons and neutrons. The atomic weight of an element is the average weight of its isotopes compared to a standard which used to be the atomic weight of oxygen, and which was considered to be 16. This 16 was really the average of the mass numbers of the isotopes of oxygen. The standard of oxygen 16 was chosen before anything was known about isotopes. For several reasons, in 1962 a new standard was selected by international agreement; the carbon isotope that has a mass number of 12 was given an atomic weight of 12. Some of the atomic weights had to be revised slightly as a result. The actual weight of an atom is something else altogether and is of no use to chemists, who are interested only in bulk matter and in the comparative weights of atoms.

A difference in arrangement

We have admitted that we do not know how the protons and neutrons are packed into the nucleus. A great deal more is known about the electron cloud outside the nucleus, yet it also challenges the imagination. How are the 92 electrons of uranium, each traveling tens of thousands of miles a second within that invisibly small system, organized to prevent a snarling of forces or an explosion?

We begin with the reminder that the negative charge on each electron acts like a buffer and repels all other electrons that approach too near. Each electron is held in its orbit by the pull of the positive charges on the nucleus. Should we imagine hydrogen's single electron so far from the nucleus that even uranium's 92 electrons could be accommodated at exactly the same distance? Could all atoms, in other words, have the same diameter? Then, the heavier the atom, the more protons there are, the more electrons the nucleus might clasp as satellites, the closer these might be to one another, though all would always be the same distance from the nucleus. But several kinds of experiments prove that atoms vary considerably in size, and therefore this model is inadmissible.

Next, it is possible to imagine hydrogen's 1 electron at a certain distance from the nucleus and helium's 2 electrons a little farther out, lithium's 3 electrons still farther out, and so on, as though we were blowing up a balloon to accommodate on its surface more and more electrons. The atoms would then increase steadily in size with their weights. Ample evidence indicates that, though the atoms have different sizes, they do not get steadily larger, but, from time to time, they actually shrink as they get heavier. Therefore this model is not good either.

The last possibility is a distribution of electrons into layers of concentric spheres. After a few electrons have accumulated at one distance from the nucleus a new sphere of several more electrons is shaped outside the first one. Thus we have an analogy of different-sized balloons, one inside the other. Each balloon's surface represents the orbiting of the electrons at that distance from the nucleus.

Until now the only energy we have talked about has been the kinetic energy of molecules and atoms, which is manifested as heat. Inside the atom there are energies related to electromagnetic fields and to radiations and to other forces about which we can only make surmises. Therefore, when we say that an electron in the atom has a certain amount of energy, we refer not just to its speed but to other energies as well. Quantum mechanics, the mathematical theory of matter and radiation, in a sense a special branch of physics founded by Niels Bohr and developed by others, defines four different measures or counts of energy for the electron which relate or fit it into the rest of the atom in such a way that no two electrons in an atom can have identical energy counts or quantum numbers. No two electrons can whirl and spin and orbit in exactly the same way inside the field of an attraction exerted on them by the same positively charged nucleus.

Since the electrons are grouped into concentric shells or spheres one inside the other, electrons in different shells will have different energies because of their different distances from the nucleus, but electrons in the same shell must be identifiably different too. For instance, all the electrons found the same distance from the nucleus do not travel in perfect circles, but one may trace out dumbbell-shaped paths and another ellipses, and there may be more complicated orbits. The direction in which they go around the nucleus will also make a difference, and so will the direction of their spin on their own axis.

All this is, in a manner of speaking, because of the wave nature of the electron. But our model of shells one outside the other does symbolize many of the facts and does not rule out many other facts that we know but cannot symbolize. Our model is extremely useful even for advanced discussion of the balance of energies inside the atom. The paper and the inked lines on a map do not resemble in the slightest the concrete highways running through a verdant or snowy countryside, but the map is invaluable for discussions of distances and directions in that countryside, and it does not deny the possibility of birds in the trees along the highway, or of fatal crashes at crossroads.

At this stage the most important energy count to consider in the atom's over-all ability to maintain its shape—though we do not assume that the atom is rigidly structured—is that which refers to the electron's distance from the nucleus, or to the particular shell it shares with other electrons. The farther an electron is from the nucleus, we say, the greater must its total energy content be. The closer an electron is to the nucleus, the greater is the influence of the positive charge on it and the less freedom the electron has. The electron farther out, therefore, has greater freedom and more free energy because less of its total energy is absorbed or neutralized by the positive field.

Examining hydrogen's single electron, we find that if somehow it obtains more energy it will move into a wider orbit—a shell farther out. If we keep exciting it, it will continue to leap into more and more distant shells, and eventually it will escape the proton's field altogether. We will be left with a separate proton and a separate electron. Any less-excited electron coming into the proton's field will click into one of the outer shells to form a hydrogen atom. If left alone, this electron will fall back from outer shells into lower ones until it reaches the innermost one, which is the electron's ground state, or lowest energy level.

As we pointed out briefly in a previous chapter, this movement out and then back toward the nucleus is not a smooth, continuous process but is made in discontinuous, instantaneous jumps from one "permitted" orbit to another. Though it may seem unremarkable now, everything about this theory stunned scientists when Bohr suggested it in 1913. It meant that energy could only be absorbed or emitted by the atom in finite little bundles—in quanta. That heat energy exists only in tiny discrete packages called quanta had first been proposed by the German physicist Max Planck at the turn of the century to account for the way in which a black body radiates heat. Five years later Albert Einstein showed that Planck's quantum concept could also explain certain properties of light and concluded that all electromagnetic radiation is a series of discontinuous packages of energy, instead of a continuous flow as pictured by the wave theory of light. But it was Bohr who conceived the brilliant idea of applying the quantum theory to the behavior of the electron within the atom. He assumed that the electron can exist only in certain specific orbits—or energy levels. If the electron absorbs a quantum of light, called a photon, it will then jump into a larger orbit or shell around the nucleus, and when the photon is emitted by the electron, it simultaneously reappears in the lower energy level, a shell nearer the nucleus.

A rough analogy is a loaded machine gun, which informs a target of its presence by the impact of its bullets. The weight and speed of each bullet is a quantum. The machine gun cannot produce an endless pipe of lead. Another analogy for the quantized energy levels within the atom is the gear mechanism of a car. The car can travel only in first, second, third, or fourth gear, never in a fraction of a gear.

Every electron in an atom can be excited to occupy higher levels of energy by hitting the atom with another suitable, energetic atom or photon. The concept of the excited electron in an atom is vital to the explanations that modern chemistry has evolved for chemical reactions.

Our model of the atom is now as follows. If a naked nucleus of any size is examined, we will find a field around it such that, when it encounters a swarm of electrons, they will fall into positions around the nucleus to form an element, and that element will be the one in the list of elements that has the same number of protons as the naked nucleus with which we started. If we pump energy into this atom it will swell, so to speak, with the electrons occupying higher levels of energy.

11

A Pattern of Atomic Structure Emerges

The arithmetic of shells and subshells

IT is interesting to note that a century ago a chemist's vague idea of what an atom might be was pretty useless to a physicist, and the latter's vague idea was useless to a chemist. The life sciences, biology, zoology, and so on, had no hard ideas about atoms at all and didn't need any. All the sciences were asking different questions about matter and trying to invent quite different structures to answer those different questions. Today the needs of all the sciences are best served by the same model of the atom, partly because the questions being asked in every science are more and more concerned with energy changes at the molecular levels of association, but largely because the wealth of information about the atom cannot be ignored. In other words, knowledge—not theory—of atomic structure is definitely changing the aims and practices of the different disciplines.

Let us now turn our attention to the electron cloud itself.

And again we first line up the elements in order of increasing protons in the nucleus, as on pages 92–93.

The first atom, hydrogen (H), has one proton and one electron. The next heavier atom, helium (He), has two protons, and its two electrons are at the same distance from the nucleus but circling with different quanta of energy. The third atom, lithium (Li), has three electrons to balance three protons. Two of these electrons take up positions similar to that of the two around the helium nucleus, but the third one cannot find a place in that shell. It has to take up a position by itself farther out, where it weaves its solitary shell around the other two. The fourth element, beryllium (Be), repeats the same arrangement: the first two electrons complete the inner shell, the third and fourth create a second shell. In the fifth element, boron (B), the fifth electron joins the two others in the outer shell.

Each new electron as we move to heavier atoms goes into this second shell until we reach the tenth element, neon (Ne), which has eight electrons in the second shell, with the usual two in the first inner shell, making ten electrons altogether, to balance the ten protons. The eleventh element, sodium (Na), has two in its first shell, and then eight in its second shell, like neon (Ne), but the eleventh electron, instead of joining those eight, begins a new, third shell at a farther distance still from the nucleus. Magnesium's (Mg) twelfth electron takes up a position in this third shell. As we proceed along the line of increasingly heavy elements, each new electron goes into the third shell until it contains eight, after which a new shell begins with potassium (K). We find that after every element which has eight electrons in its outermost shell the next element begins a new shell with its new electron, instead of making it the ninth in the outermost shell. Thus neon (Ne), argon (Ar), krypton (Kr), xenon (Xe), radon (Ra) all have eight electrons in their outermost shells, and the elements immediately following them, sodium (Na), potassium (K), rubidium (Rb), cesium (Cs), and franconium (Fr), each have one electron in a new outermost shell, outside the one with eight in it. The exceptions are helium (He), with only two in its shell, and lithium (Li), which follows it with a new shell outside that.

By the time we reach uranium (U) we find that its 92 electrons are distributed into seven concentric shells, one inside the other, with the nucleus containing its 92 protons and 146 neutrons at the center. The seven successive shells have been assigned the letters K, L, M, N, O, P, Q, or the numbers 1, 2, 3, 4, 5, 6, 7, starting with K or 1 for the innermost shell.

Each shell must be imagined as a region or band in which a maximum number of electrons can exist without colliding. The

At. No.	Name of Element	Symbol	At. No.	Name of Element	Symbol
1	Hydrogen	H	27	Cobalt	Co
2	*Helium*	*He*	28	Nickel	Ni
3	Lithium	Li	29	Copper	Cu
4	Beryllium	Be	30	Zinc	Zn
5	Boron	B	31	Gallium	Ga
6	Carbon	C	32	Germanium	Ge
7	Nitrogen	N	33	Arsenic	As
8	Oxygen	O	34	Selenium	Se
9	Fluorine	F	35	Bromine	Br
10	*Neon*	*Ne*	36	*Krypton*	*Kr*
11	Sodium	Na	37	Rubidium	Rb
12	Magnesium	Mg	38	Strontium	Sr
13	Aluminum	Al	39	Yttrium	Y
14	Silicon	Si	40	Zirconium	Zr
15	Phosphorus	P	41	Niobium	Nb
16	Sulfur	S	42	Molybdenum	Mo
17	Chlorine	Cl	43	Technetium	Tc
18	*Argon*	*Ar*	44	Ruthenium	Ru
19	Potassium	K	45	Rhodium	Rh
20	Calcium	Ca	46	Palladium	Pd
21	Scandium	Sc	47	Silver	Ag
22	Titanium	Ti	48	Cadmium	Cd
23	Vanadium	V	49	Indium	In
24	Chromium	Cr	50	Tin	Sn
25	Manganese	Mn	51	Antimony	Sb
26	Iron	Fe	52	Tellurium	Te

LIST OF CHEMICAL ELEMENTS in the order of increasing atomic number. The atomic number indicates the number of protons in the nucleus of the atom and, hence, since the atom is

At. No.	Name of Element	Symbol	At. No.	Name of Element	Symbol
53	Iodine	I	79	Gold	Au
54	*Xenon*	*Xe*	80	Mercury	Hg
55	Cesium	Cs	81	Thallium	Tl
56	Barium	Ba	82	Lead	Pb
57	Lanthanum	La	83	Bismuth	Bi
58	Cerium	Ce	84	Polonium	Po
59	Praseodymium	Pr	85	Astatine	At
60	Neodymium	Nd	86	*Radon*	*Rn*
61	Promethium	Pm	87	Francium	Fr
62	Samarium	Sm	88	Radium	Ra
63	Europium	Eu	89	Actinium	Ac
64	Gadolinium	Gd	90	Thorium	Th
65	Terbium	Tb	91	Protactinium	Pa
66	Dysprosium	Dy	92	Uranium	U
67	Holmium	Ho	93	Neptunium	Np
68	Erbium	Er	94	Plutonium	Pu
69	Thulium	Tm	95	Americium	Am
70	Ytterbium	Yb	96	Curium	Cm
71	Lutetium	Lu	97	Berkelium	Bk
72	Hafnium	Hf	98	Californium	Cf
73	Tantalum	Ta	99	Einsteinium	Es
74	Tungsten	W	100	Fermium	Fm
75	Rhenium	Re	101	Mendelevium	Md
76	Osmium	Os	102	Nobelium	No
77	Iridium	Ir	103	Lawrencium	Lw
78	Platinum	Pt			

neutral, also the number of its electrons. Those elements appearing in italics in the list, have a stable configuration of eight electrons in their outermost shell, except for helium, which has two.

larger the shell, the more electrons it can accommodate. This means that in every shell, whether or not it contains electrons, the potential exists for containing more than any shell within it can contain. Since each shell represents a band of energies, and since the position of any electron in any shell can be defined by its quantum numbers, and since no two electrons can have the same quantum numbers, the relative positions of the electrons within any one shell must also be indicated. This is done by imagining subshells and giving them the letters *s*, *p*, *d*, *f*, *g*, etc. Every shell, whatever the atom's number, and however many electrons it has, is considered to have all these subshell positions. Shell K has positions for its maximum of two electrons in its *s*, *p*, *d*, and *f* subshells, even though in the atom's normal state its two electrons always occupy only subshell *s*. Shell L, the second shell, has subshells 2*s*, 2*p*, 2*d*, and 2*f*, and so on. Subshells exist with the letters *s*, *p*, *d*, *f* for every shell, with the number of the shell preceding them.

There is still one more item from quantum mechanics which adds to our knowledge. Every electron has a partner—that is, the energies of any electron are meshed a little more intimately with one other electron than with all the others. Electrons pair together in the subshells into what are called orbitals. Obviously, an odd number of electrons must leave one by itself, unpaired, outermost. But in the next atom this loner is usually paired with the new electron that has been added. That is to say, the energies of electrons in these shells are such that one can recognize an electron that is not matched by another to make a pair. By definition any orbital can contain only one pair. An orbital therefore may have one or two electrons in it, but never more than a pair.

We are interested in the actual number of electrons each neutral atom arranges around its nucleus in its normal state; therefore we label the subshells that electrons can occupy around the nucleus, rather than the electrons. We can refer to an electron in the second orbital of 3*f* subshell.

In every atom after hydrogen the K shell or shell no. 1 always contains two electrons in the 1*s* subshell. The maximum number of electrons possible in the L shell, or shell no. 2, is eight. The first two are in the 2s subshell. The next two are in the 2p subshell; the next two, and the next two, are all in the 2p subshell, occupying higher and higher levels of energy in their orbitals. The electrons in any atom's completed M shell are found thus: two in the 3s subshell, six in the 3p subshell, and ten in the 3d subshell.

The maximum number of electrons that a shell can contain is given by a surprisingly simple formula: $2n^2$, where n is the number of the shell. Shell K, no. 1, can therefore never hold more than 2×1^2, which is 2. Shell L contains 2×2^2, or 8. Shell M can accommodate in its orbitals a total of 2×3^2, or 18 electrons. Shell N has 32, shell O would have 50, and so on.

But no matter how many electrons a shell contains when it is completed, research reveals that *there are never more than two electrons in K and never more than eight electrons in any other shell while it is the outermost shell.* As soon as a shell—except K—has acquired eight electrons the next element begins a new outer shell.

A periodic chart of the atoms by outermost shells

If we now chop up the list of elements, using the sequence on pages 92 and 93 to aid the imagination, in such a way that each section after helium begins with an element that has a single electron in a fresh outermost shell and ends with an element having eight electrons in its outermost shell, and if we move these sections under one another, placing all the elements with one electron outermost under one another, and those with two electrons outermost under one another, and so on, we will have arranged our list into a periodic table of the elements. Each of the seven horizontal sections is called a period, and each vertical column is called a group. The groups can be numbered according to the number of electrons the elements in the group have *outermost*. Since an outermost shell never contains more than eight electrons there ought to be eight groups, and the fact that there are eighteen elements in the fourth and also in the fifth periods and thirty-two elements in the sixth will be explained shortly.

The thing to note is that this periodic arrangement of the elements allows us to see any atom in two ways. First, it belongs to a horizontal period whose members always have the same number of shells but a different number of electrons within the shells. And second, each atom also belongs to a vertical group whose members have a different number of shells but always the same number of electrons outermost, or in the shells being filled. The man-made elements fit into the table following uranium in the seventh period, which is incomplete.

Once again we have extracted, with scientific tools, observations that reveal a beautifully balanced order in nature, the details having been collected here and there for different reasons in different times, without any preconception of that order to guide the research. The basic electronic structure of all atoms is the same, and any atom differs from all others by a mathematical count of its parts, which trace different shapes to balance the forces holding the parts together against forces pushing them apart. No one fifty years ago had any inkling that this was the case.

Of course, as soon as we enunciate a lovely simplification we must begin to qualify it as though it were only a rough rule of thumb instead of the whole truth. Revision is the fate of all scientific laws. If to some cosmic observer the universe appears to be a unified whole and not a tangle of unrelated events, not a jumble of unpredictable accidents, then local patterns and harmonies must reflect somehow the vast happenings of which they are a part. Every crevice and crumb must fit its reality into the universal movement somehow. It is the discovery of local rules or laws that allows us to believe that the universe is a unified whole, even though proof of this is beyond our ken. Since all the parts of any system that we discover are also in contact with the whole rest and thus not complete in themselves, we cannot expect to find perfect systems within the universe. If anything is perfect it can only be the total universe.

This idea is seldom if ever proposed in science books, though

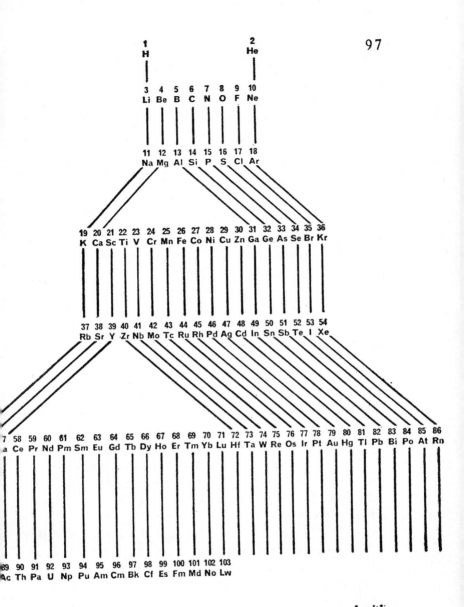

A PERIODIC CHART OF THE ELEMENTS arranged with those elements having a single electron in their outermost shells at the left and those with their outermost shells complete at the right end of each period, or horizontal row.

to any thoughtful scientist it is an obvious concept and it is what many philosophers and mystics have been saying from the beginnings of recorded history. But our search for the mechanisms of nature would at once become vain if we reintroduced the ancient belief that certain laws of nature are finer and more meaningful, more perfect in the over-all sense of things, than other laws. There cannot be an ideal electron or an ideal galaxy consisting of ideal atoms.

The gradual working out of the periodic aspect, meaning the repetitious structuring, of the electron cloud in the atoms of the various elements, illustrates how our yearning to understand nature in order to control it succeeds if we proceed exclusively with laboratory facts, which at bottom are numbers, and are never directed by social principles or nonscientific goals.

Let us resume our examination of the atoms from which are constructed the brains that discovered the atom and that speculate on the future of man.

The conventional form, called the long form, of the periodic table of the elements (pp. 100 and 101) reveals how each of the shells is built up in the elements of the successive periods.

In the first period there are only hydrogen (H) and helium (He). With helium the first shell, K, is completed.

Lithium (Li) begins the second period with two electrons in its K shell and a single electron in shell L. This period ends with neon (Ne), which has a complete L shell of eight electrons.

Sodium (Na) starts the third period—the third shell, M—with an electron arrangement of two in its K shell, eight in its L shell, and one in its M shell. Argon, with eight electrons in its M shell, ends this period, even though M is not yet complete, having a total capacity of eighteen electrons.

The fourth period begins with shell N, and the first atom with an electron in N is potassium (K), whose electron configuration is 2, 8, 8, 1, a total of nineteen, this being the nineteenth element. Calcium (Ca) is the twentieth element, with twenty protons and its twenty electrons arranged 2, 8, 8, 2. The twenty-first element, scandium (Sc), might be expected to add a third electron to the N shell, but instead this third electron slips down into the un-completed M shell to produce a configuration of 2, 8, 9, 2. The twenty-second element, titanium (Ti), puts its twenty-second elec-tron into M also: 2, 8, 10, 2. And as we continue along this fourth period we find the following elements all adding their new elec-trons to the inner M shell while the outer N maintains itself at two electrons, sometimes only one. Only when M is completed with its full load of eighteen electrons in element no. 30, zinc (Zn), does N resume its filling up. When N has eight, in krypton (Kr), with an electron configuration of 2, 8, 18, 8, the period ends. But the N shell is still not complete.

The fifth period begins with rubidium (Rb), with 2, 8, 18, 8, 1, the single electron starting the new shell O, the fifth shell. Stron-tium (Sr) adds a second electron to the O shell but then yttrium (Y) puts its new electron into the still uncompleted N shell— 2, 8, 18, 9, 2. The N-shell is now built up, while the outermost O shell keeps two or sometimes one electron in its lowest orbital, until cadmium (Cd), when N has eighteen, after which electrons are added to shell O. Xenon (Xe) has eight in the O shell, and the period ends without having completed shell N. The next period follows the same scheme of filling up inner shells while the outermost one marks time with two or one electrons in it.

This quite unexpected process of completing the larger shells only when they have become inner shells, thus allowing no shell to remain outermost with more than eight electrons, can be ex-plained in terms of energy balances.

A shell represents a level of energy created in the space sur-rounding the positively charged nucleus. The levels of energy in-crease with increasing distance away from the nucleus—that is, an electron close to the nucleus has a total energy which is less than that of an electron farther away from it. Each new electron added to compensate for each new proton in the nucleus fills that available position which has the lowest energy level. In each shell the band of subshells increases in energy potential outward from *s*

The Periodic Table

Number of electrons
in each shell at the
start of each period

$\vdash$ **s subshell** $\dashv$
being filled

K	L	M	N	O	P	Q
I						
2	I					
2	8	I				
2	8	8	I			
2	8	18	8	I		
2	8	18	18	8	I	
2	8	18	32	18	8	I

	IA	IIA	IIIB	IVB	VB.	VIB	
Period 1	1 H						
Period 2	3 Li	4 Be					
Period 3	11 Na	12 Mg		d subs			
Period 4	19 K	20 Ca	21 Sc	22 Ti	23 V	24 Cr	
Period 5	37 Rb	38 Sr	39 Y	40 Zr	41 Nb	42 Mo	
Period 6	55 Cs	56 Ba	57–71 La*	72 Hf	73 Ta	74 W	
Period 7	87 Fr	88 Ra	89–103 Ac⁻⁺	104	105	106	

IN THIS PERIODIC TABLE of natural and synthetic elements the vertical groups represent elements with the same number of electrons in their outermost shells, the A and B subgroups differing from each other with respect to the number of electrons in their two, or three, outermost shells. The synthetic elements are indicated by cross-hatching. Period 7 is incomplete, elements 104 to 118 being as yet undiscovered.

*
$\vdash$ **LANTHANIDE SER**

58	59	60	
Ce	Pr	Nd	

-¦-
$\vdash$ **ACTINIDE SERIES**

90	91	92	
Th	Pa	U	

# of Chemical Elements

|—— P subshell being filled ——|

								INERT GASES
			IIIA	IVA	VA	VIA	VIIA	2 He
			5 B	6 C	7 N	8 O	9 F	10 Ne

g filled ————|

— VIII —	IB	IIB	13 Al	14 Si	15 P	16 S	17 Cl	18 Ar
27 Co	28 Ni	29 Cu	30 Zn	31 Ga	32 Ge	33 As	34 Se	35 Br	36 Kr
45 Rh	46 Pd	47 Ag	48 Cd	49 In	50 Sn	51 Sb	52 Te	53 I	54 Xe
77 Ir	78 Pt	79 Au	80 Hg	81 Tl	82 Pb	83 Bi	84 Po	85 At	86 Rn
109	110	111	112	113	114	115	116	117	118

subshell being filled ————

63 Eu	64 Gd	65 Tb	66 Dy	67 Ho	68 Er	69 Tm	70 Yb	71 Lu

subshell being filled ————

95 Am	96 Cm	97 Bk	98 Cf	99 Es	100 Fm	101 Md	102 No	103 Lw

to p to d to f, but a complication arises in that the total energy span of the larger shells is such that bands of subshells overlap. For example, the inner subshell of shell N is closer to the nucleus than the outermost subshell of M. Hence, in potassium, the nineteenth electron finds that the empty place with the lowest energy level is not in the incomplete outermost subshell of the M shell, but in the inner subshell of N. The twentieth electron of calcium (Ca) also goes into the inner N subshell, which has room for two electrons. But the lowest available energy for the twenty-first electron in scandium (Sc) is back again in the outermost subshell of the M shell.

Considering all that we know about the electron's wave properties and its speed when viewed as a particle, plainly we cannot possibly create a model in which this overlapping is represented with any clarity. According to our system of notation, an electron finds that the first energy level in $3d$ is lower than the first energy level in $4p$. An electron always goes to the position of lowest energy available to it, so that there are never any "holes" of energy, never any flaws in the sequence of the arrangement.

When a shell has eight electrons in it the configuration is so stable that the larger nuclei of the next few elements in the table cannot pull the new electrons down into the shell. Even though the pull of the nucleus is stronger it has to hold on to two new electrons out beyond the eight. After that the nucleus is strong enough to break up the arrangement and pull electrons into the underlying shell. Once the arrangement of eight is broken, all the new electrons pile into that shell until it reaches another, though considerably less stable, configuration when it is completed with eighteen.

Let us now take a more knowledgeable look at the periodic arrangement in the table on pages 100–101. Note that there are A and B subgroups. This will be clarified later. At the moment the A groups interest us.

Hydrogen and helium comprise the first period.

The second period has eight elements and it establishes the number for each group according to the number of electrons in the

L shell. The third period falls into place with a similar arrangement of electrons in the M shell. The fourth and subsequent periods have to be stretched out with the B subgroups, but they do fit, and a certain amount of study will clarify the mechanical spotting of the elements in their proper places. In each vertical group, the elements have the number of electrons outermost that the group number indicates. By looking up the group number of any element in the first three periods, and in all the A groups, we can instantly say how many electrons it has in its *outermost* shell. This statement will have to be modified later, but at the moment it is a fair generality to make.

In all this discussion of the structure of the atom we have not once mentioned those properties of the elements by which our senses recognize them. Gold is not like phosphorus; oxygen is not like aluminum; nitrogen is not like calcium. Anyone at all can describe in great detail the different properties of the elements, and anyone at all can find some similarities among them. Gold is a metal and so is silver, and they have certain properties, such as malleability, that other metals have too. Sulfur is a yellow solid and oxygen a colorless gas, yet they combine with other elements in a very similar way. Nitrogen, the gas we breathe in and out without being affected, and arsenic, which is a deadly poison, are remarkably similar in the compounds that they form with oxygen.

It is like and unlike physical and chemical properties that define substances for us and that defined them for chemists in the past, and we now focus our attention on them.

12

Similar Properties Mean Similar Structures

Properties, like and unlike

IT is a cliché that the college graduate who has not taken any science courses knows less about the universe than Aristotle did. We can reason as well as, but no better than, the Greeks concerning the nature of matter if we use only our natural sense impressions, and today our senses are held just as fast by the brute bulk appearance of matter and its infinite variety as were man's senses two thousand years ago. Fundamentally our scientific instruments are built as extensions to the mind's ability to recognize like and unlike properties. But it is still reason that transforms the information our inventions harvest into the knowledge and the power of our age. Even an electronic computer works by obtaining a physical impression—i.e., of holes in a card—then evaluating the impression and sorting its memory for matching answers: yes, no; like, unlike.

From our first breath to our last, resemblances, recurrences, echoes, similarities give us an illusion of order in the universe and of the possibility of planning for the future. The resemblances, and other kinds of association, need not be obvious or simple or easy to define. Roots, bark, trunk, leaves, twigs, unfurling of buds, and withering of foliage are properties of every tree, though no two trees are alike. A piece of plastic painted to look like a maple leaf might fool even our sense of touch, yet it is not a leaf.

But the question of what are genuine properties of objects and what are properties of the mind observing the objects has plagued man from the beginning of philosophy. Today neurologists, psychologists, biochemists, physicists are pooling their imaginations to determine whether the brain is a blank slate at birth and all knowledge is learned, or whether some knowledge is innate. For instance, does our impression of color exist in the stimulated eye, or somewhere in the brain behind the eye? The smallness of an object at a distance does not confuse us as to its real size; is this the kind of information that depends on learning or is it inherent at birth? After all, it's never the object itself that enters our brain.

Therefore "property" is a word which, like "aspect" or "appearance" or "the nature of things," slowly crumbles when we put it to rigorous examination, whether philosophic or scientific. Is beauty in the eye of the beholder? That is easy to answer—yes. But beauty is not a property in the scientific eye. On the other hand, without such generalized meanings lacing our speech we could hardly communicate anything. The word "property" is one which implies all the words with which any experience can be described, including the sensation of fitness, perhaps, or harmony, symmetry, balance, aesthetic tensions: beauty. Scientists use the phrase "a beautiful experiment."

Still, here, "property" refers exclusively to the nature of objects, not of the mind, and any one kind of property can be applied to many kinds of substance. Jewels are distinguishable from common stones by their glitter and polish, color, hardness, translucency, crystal shape, and by what kind of liquid they dissolve in. There are many substances that are saltlike in taste, that dissolve in water, that are crystalline, and that are produced in ways similar to the way table salt can be produced. There are many liquids that have a property we call acidic. There are inflammable gases, and there are poisons of all sorts which resemble one another only in that if a man swallows them he dies.

One of the most striking sets of resemblances belongs to the elements we call metals. They are hard, with surface areas that can be polished to mirror-like smoothness; they can be hammered into sheets and drawn into wires, their tensile strength is high; they react with acids to produce hydrogen; they can be melted down and mixed with other melted metals into a mixture that hardens into alloys. In other words, the concept of metal grows out of the properties that a number of elements share. The members of a group are much more like one another than they are like members of other groups. But can these metallic properties be found in other substances? Yes. Other things can be polished into mirrors too. And other things have tensile strength. When glass is warmed, it too can be rolled and drawn out fine. Some substances that are not metals conduct electricity quite well. But a metal is an element that shows more of these properties in a certain range or degree than any other element does, and this is why standards are necessary which will provide us with numerical values for comparison, such as melting points.

At the end of the first quarter of the nineteenth century, when ideas of elements began to shape into the first premise—really only a conviction—that an element is a natural and tangible substance which cannot be broken down into other substances, the German chemist Johann Döbereiner collected and studied groups of elements that not only resembled one another physically and chemically but did so with a mathematical ratio to their resemblance. The groups always consisted of three elements, such as gold, silver, and copper, or iron, cobalt, and nickel, or chlorine, bromine, and iodine. The middle element of a triad always had properties midway between those of the other two. Some of these groupings had been known for a long time; others he dug out of the growing list of new elements that were being discovered. The prevalence of the magic number three no doubt made other scientists suspicious that Döbereiner was dabbling in residual alchemy, but for most chemists the triads were just curiosities of nature without significance. A few elements were bound to resemble one another, ran the reasoning, but to suggest that there might be a mysterious order to the endless variations of matter and its qualities was not scientific. It was putting the cart before the horse, a return to deductive reasoning: it was necessary to collect enough evidence, first, to support a theory that, without the evidence, cannot be formulated.

Still, some men were not made uneasy by the ancient hope of discovering an unsuspected orderliness in this matter of elements. In the 1860s the British chemist John A. R. Newlands succeeded in exploring the idea further when many more elements had been isolated and studied. He discovered that if he drew up a list using the comparative atomic weights, starting with hydrogen, the lightest, in some cases each element resembled another eight places away from it on his list. That is, elements 3, 11, 19 resembled one another more than they resembled elements 4, 12, 20, while these latter resembled one another strikingly.

To refresh the reader's memory: some of the physical properties of an element, or a compound, or even a mixture, are its density, melting and boiling points, vapor pressure, hardness, color, taste, smell, tensile strength, compressibility, electrical conductivity, malleability, crystal shape, angle of light refraction, solubility in various liquids, the temperature at which it burns.

Chemical properties are a list of the elements and compounds that a substance reacts with chemically, together with information on the nature of those reactions: how much heat is evolved or absorbed, the speed with which the reaction proceeds, and so on.

When Newlands spoke of a physical resemblance among three or four elements he meant that their melting points, density, hardness, et cetera, were more or less on the same order. When he spoke of chemical resemblance he meant that the compounds that members formed with other elements resembled one another too. For instance, one group was acid-forming.

But these octaves flabbergasted chemists even more than the triads had, since there was considerably more information involved here. Was there any more significance to the octaves than to the triads, or were they just coincidental too? There did not seem to be any useful purpose in such a relationship among the elements. It explained nothing and could not itself be explained.

The adventurous-minded, however, accepted the possibility that the nature of the elements—atoms were still a matter for cautious discussion—reflected a harmonious rhythm, as do the octaves of music. Music is the highest art and the only one mathematically structured. Each note is a multiple of some basic wavelength of sound and reappears at intervals of eight on a scale—in the Western World until this century—that can contain as many octaves as desired. One could chose five or fifty basic wavelengths, but the principle was the same. Every octave repeated the same notes higher or lower on the scale. The idea that music somehow is part of the cosmic scheme is as old as the Pythagorean idea of perfect heavenly spheres revolving musically in perfect harmony.

But the most intensive search turned up only a few octaves and only among the lighter elements. The melody turned to cacophony among the heavier elements.

By that time, the second half of the nineteenth century, much of the laboratory technique, equipment, reagents, methods of reporting scales of properties, nomenclature, and jargon had become standardized at international conferences. Participators in these affairs were noisy, impatient, quarrelsome, often dull, often pompous, sometimes brilliant, and a variety of egos were refreshed or crushed, but they were scientists intensely dedicated to the sharing of knowledge and to the improvement of communication. For the first time in history a body of public knowledge was infinitely more important than any one man's discovery, or any organization's power. It was not only that international get-togethers and regular publications and visiting lecturers spread scientific information, it was that the new world of science insisted on getting everything known to everyone. One of the prime problems in science today is that new information is of flood proportions and researchers cannot hope to keep abreast of even the most important developments.

The increasing urgency with which the nineteenth century sought to systematize scientific data was due at least partly to the increasingly important role science was playing in medicine, agriculture, manufacturing, mining, mental and social hygiene. Thus, as a matter of course in the correlating of data for practical purposes, it was obvious to most chemists that many elements did resemble other elements and that it was convenient to talk of groups of elements without knowing theoretically why they were similar. The hope that the elements really were constructed according to some basic principle, even if it was not like that of music, kept some men industriously arranging and rearranging them in various ways as they hunted for patterns.

Mendeleev's Periodic Law

Finally in 1869 the Russian chemist Dmitri Mendeleev proposed that the properties of the elements were a periodic function of their atomic weights. A year later a German, Lothar Meyer, published a similar thesis, developed independently, but Mendeleev is given the palm.

He wrote out the elements in order of their relative atomic weights, from lightest to heaviest, starting with hydrogen, and he showed that certain properties—the ability to conduct electricity, for example—reappeared at regular intervals along the line, as the properties of specific elements. To put it more generally, the properties of any element resemble the properties of several other elements spotted at fairly regular intervals through the list of elements.

Two hydrogen atoms combine with one oxygen atom to form a molecule of water; one atom of oxygen also combines with two atoms of lithium, and also with two atoms of sodium, and also with two atoms of potassium. We can say that the chemical property of combining in twos with one atom of oxygen is shared by hydrogen, lithium, sodium, potassium; further, these oxides resemble one another in specific ways. For example, when dissolved in water they all form bases.

Fluorine, chlorine, bromine, and iodine resemble one another not at all at first glance. Fluorine is a pale yellow gas, chlorine is a yellow-greenish gas, bromine is a brown liquid with a red vapor, and iodine is a dark solid with a purple vapor. But they all combine in much the same way with hydrogen, for example, to form compounds which, mixed with water, produce acids that are also similar.

Mendeleev, having charted the recurring properties on his list of the elements that were known to him, cut up the list into sections, or periods, in such a way that each section began with a strikingly similar element—one of the alkali metals. He left out hydrogen because it seemed to resemble nothing else. Then he moved successive sections, or periods, under one another so that all the alkali metals now formed a vertical column. In each of the other vertical columns the elements were also similar to one another. For example, chlorine, bromine, iodine, and fluorine were all in one column, or group.

The correlation was not beautifully, mathematically precise but the majority of the known elements quite clearly did have a great deal in common with the others in the same vertical group, since Mendeleev had the foresight to realize that a number of elements had not yet been discovered and that blank places had to be left for them in his table.

Mendeleev was not presenting the world with factual data it did not know; he had not discovered any new elements or new information about the seventy-odd known elements, or even any original technique, and he had no new theory to explain anything. All he had done was line up the elements in rows and columns and point out what anyone could see at a glance, once it was pointed out. Despite all this, his chart was one of the great historical shockers to scientists and probably the most important single contribution to chemistry in the nineteenth century. It just seemed, at the time, incredible that this wonderful arrangement could exist without anyone's having noticed it long ago.

Most scientists rejected it outright on the ground that Mendeleev's statement, that the properties of the elements were a periodic function of their atomic weights, was not theoretical but simply a statement of what appeared to be amazing coincidences. Some scientists even denied seeing the coincidences and accused Mendeleev of somehow juggling with facts to throw the spotlight of the world's opinion on him at the international conference where he introduced his chart. But even those scientists who were impressed and who were convinced it could not be all coincidence agreed that because Mendeleev had no theory to explain his periodic law it could not be called a scientific contribution. It was too wild in its implications. Not even the adherents of the idea that the different atoms were constructed out of hydrogen had dreamed of such harmonies.

The first obviously striking aspect of his periodic arrangement is that although the properties, such as hardness of a certain magnitude, change from element to element along a horizontal line, or period, they are faithfully repeated in the next period. Therefore the change from group to group cannot be utterly random but must be directional in some way. There must be a pattern of change in the very substance of the elements, in the atoms themselves.

Mendeleev used the atomic weights, which are comparative, with hydrogen being 1, as the foundation. The properties not only varied with the atomic weights, which is the same as saying that each element had its own weight and its own properties, but varied periodically with the weights. That is, the recurrence of properties is regular if the elements are in order of increasing atomic weights.

In Mendeleev's chart, the properties of any element are the average of the properties of elements on either side of it in the same period, and the average of the properties of elements above and below it in the group. In other words, the chart proves that the elements are somehow related. It indicates, for example, that the properties of phosphorus (P) are midway between the properties of its horizontal neighbors, silicon (Si) and sulfur (S), and are midway between its vertical neighbors, nitrogen (N) and arsenic (As). No matter how different silicon is from sulfur, phosphorus, arsenic, and nitrogen, the difference is nevertheless an orderly one. By taking the averages of the four elements, the properties of silicon can be calculated quite closely with simple arithmetic.

Thus, not only are there similarities in each group but there are trends along each period.

On the basis of these recognitions and without a whisper of theory to aid him, Mendeleev became convinced that the periodicity he had uncovered was a key to some fundamental nature of matter. The elements from which the universe was made were structured in a rhythmic or periodic fashion and, with a boldness that accompanies all great discoveries, he ignored certain facts in order to improve his table. When the properties of an element did not fit the group into which it fell by virtue of its place in the line-up of atomic weights, he simply moved it out of its sequence and into the group it fitted best. He switched several elements, putting a lighter one after a heavier one. Sometimes, moving an element out of its place to another, more suitable group left a vacant place in the chart which he could not fill. There were, it seemed, no elements for certain places in the chart. Not only did Mendeleev predict that those vacant positions belonged to still undiscovered elements, but he also described what these unknown elements would be like when they were discovered and urged chemists to hunt for them.

For instance, he predicted that the unknown element following gallium (Ga) ought to be a silvery metal with a certain specific gravity, a certain melting point, and so on. These figures he obtained by averaging the properties of the elements around the vacancy. A short while later the element was discovered and called germanium (Ge), and it did indeed have properties almost precisely as Mendeleev had predicted. To this day it seems strange that it was not named after him. Soon several more elements were discovered for the holes in his chart, and the scientific world agreed that Mendeleev's periodic chart was based on a natural law.

There was, however, one group of elements for which Mendeleev left no room in his table—the Group O elements, called the inert, or noble, elements, none of which were discovered until the end of the nineteenth century. This group consists of helium, neon, argon, krypton, and xenon, which are gases at ordinary temperatures and are very similar in almost every way. They all have very low boiling points, that is, they refuse to give up their heats of fusion easily, and until 1962 no compounds of them were known. These elements refuse to combine with any other atoms except under unusual conditions. No other elements are nearly as difficult to force into combinations. Thus they resemble one another far more than they resemble any other element because of their inert nature.

When all the naturally occurring elements were discovered, their total number was 88, although uranium, the heaviest, was element 92. Thus four gaps remained in the table until atomic structure was understood well enough to allow a deliberate creation of these four in nuclear research. They are so highly radioactive that they had vanished from the earth. Two are heavier than lead, atomic number 82, but all elements heavier than lead are radioactive and they survive on this planet because their rate of disintegration is slow enough. Elements 93 to 103 have been created in nuclear research but they, like the missing four, have comparatively brief radioactive lives.

We must emphasize that at first most serious scientists considered Mendeleev's periodic table a fascinating piece of un-science because no one could propose a theory to explain it. Furthermore, Mendeleev had had to fiddle with his own rules in order to make the elements fit into the right groups. Such objections may seem pedantic today in the face of the astonishing regularity everyone saw in the chart, but the objectors were using sound reasoning for their day, when "average" and "probable" were weak words compared to mathematical precisions.

Properties of the elements depend on atomic structure

Now let us compare an updated and expanded version of Mendeleev's chart of the elements arranged on the basis of their physical and chemical properties (pages 114–115) with the table on pages 100 and 101, constructed on the basis of electron configurations within the atom. The two charts match exactly. The periodicity of the properties follows precisely the periodicity of the electron shells. We can therefore say at once that the chemical and physical properties of the elements somehow relate directly to the electron and nuclear structure of the atom.

After all, this discovery is what we should expect. Only the

GROUP	I		II		III		IV	
	A	B	A	B	B	A	B	
Period 1	1 Hydrogen (1.00797)							
Period 2	3 Lithium (6.939)		4 Beryllium (9.012)		5 Boron (10.81)		C (1	
Period 3	11 Sodium (22.990)		12 Magnesium (24.31)		13 Aluminum (26.98)		S (	
Period 4	19 Potassium (39.102)		20 Calcium (40.08)		21 Scandium (44.96)		22 Titanium (47.90)	
		29 Copper (63.54)		30 Zinc (65.37)		31 Gallium (69.72)		Germa (
Period 5	37 Rubidium (85.47)		38 Strontium (87.62)		39 Yttrium (88.905)		40 Zirconium (91.22)	
		47 Silver (107.870)		48 Cadmium (112.40)		49 Indium (114.82)		(1
Period 6	55 Cesium (132.905)		56 Barium (137.34)		57-71 Lanthanide series*		72 Hafnium (178.49)	
		79 Gold (196.97)		80 Mercury (200.59)		81 Thallium (204.37)		(2
Period 7	87 Francium (223)		88 Radium (226)		89— Actinide series†			

SHORT-FORM PERIODIC TABLE based on Mendeleev's chart. The atomic weight in accordance with the most recent authoritative reports is given in parentheses below the name of each element.

*LANTHANIDE SERIES

57 Lanthanium (138.91)	58 Cerium (140.12)
65 Terbium (158.92)	66 Dysprosiu (162.50)

†ACTINIDE SERIES

89 Actinium (227)	90 Thorium (232.04)
97 Berkelium (247)	98 Californiu (251)

	VI		VII		VIII			O
A	**B**		**A**	**B**		**A**		
								2 Helium (4.003)
7 itrogen (14.007)	8 Oxygen (15.9994)		9 Fluorine (19.00)					10 Neon (20.183)
15 phorus (30.974)	16 Sulfur (32.064)		17 Chlorine (35.453)					18 Argon (39.948)
ium	24 Chromium (52.00)		25 Manganese (54.94)		26 Iron (55.85)	27 Cobalt (58.93)	28 Nickel (58.71)	
33 Arsenic (74.92)	34 Selenium (78.96)		35 Bromine (79.909)					36 Krypton (83.80)
m	42 Molybdenum (95.94)		43 Technetium (98)		44 Ruthenium (101.1)	45 Rhodium (102.905)	46 Palladium (106.4)	
51 timony (121.75)	52 Tellurium (127.60)		53 Iodine (126.90)					54 Xenon (131.30)
um	74 Tungsten (183.85)		75 Rhenium (186.2)		76 Osmium (190.2)	77 Iridium (192.2)	78 Platinum (195.09)	
83 ismuth (208.98)	84 Polonium (210)		85 Astatine (210)					86 Radon (222)

dymium	60 Neodymium (144.24)	61 Promethium (147)	62 Samarium (150.35)	63 Europium (152.0)	64 Gadolinium (157.25)
	68 Erbium (167.26)	69 Thulium (168.93)	70 Ytterbium (173.04)	71 Lutetium (174.97)	

tinium	92 Uranium (238.03)	93 Neptunium (237)	94 Plutonium (242)	95 Americium (243)	96 Curium (247)
nium	100 Fermium (253)	101 Mendelevium (256)	102 Nobelium (254)	103 Lawrencian (257)	

atom's own structure can account for its properties. The only basic difference between the two tables is that in the 1860s Mendeleev used the atomic weights to line up the elements, while now we use the number of protons in the nucleus.

His table swept away a universe of independently constructed atoms and introduced a universe of atoms harmoniously related to one another, though he didn't know how. We do know how, today.

However, the method of chemical combination could not be perceived in his periodic table. Sulfur atoms were presumably like minuscule crumbs of powdery yellow sulfur which burns in oxygen with a blue flame to produce thick clouds of acrid smoke. Copper atoms were presumably hard, lustrous, and reddish, and had the potential to conduct electricity. Actually, we know that the solid and liquid states of atoms have properties pertaining to those states, and thus an atom's solitary personality is not like its personality in large quantities. But let the argument stand: it is chemically true that when a copper atom and a sulfur atom combine they form copper sulfide, which in the bulk is black and granular and neither metallic nor sulfurous. This combination of atoms into a compound can be broken up. The atoms can be separated from their molecular bonds by a series of reactions that in the end release the red copper and the yellow sulfur. The mechanism for all this cannot be discovered just by studying the modern periodic table, but the table guides us in the interpretation of additional information.

Isotopes

Before we pick up again those electronic configurations in the light of our knowledge that they are related to the ordinary properties of matter—that they therefore actually determine what those properties are—we must emphasize certain facts and reintroduce others that explain one of the discrepancies Mendeleev could not understand.

We already know that every element, except common hydrogen, has neutrons in its nucleus, as well as protons, and that each neutron has a mass of approximately 1. The mass of any atom is simply the sum of its neutrons and protons, electrons being too light to count, and the comparative weights should therefore be whole numbers. Yet very few elements have atomic weights that are round numbers. Most atomic weights contain fractions. Chlorine (Cl) 35.453, argon (Ar) is 39.948, silver (Ag) 107.870, lead (Pb) 207.19, and so on. Since we know that chlorine contains exactly 17 protons and these have a mass of 17, how can the neutrons weigh 18.453? The explanation was put together here and there, mostly in physics, and we have already referred to it briefly when isotopes were mentioned earlier in the book.

Almost every element is made up of a mixture of atoms, all of which have the same number of protons and electrons but a different number of neutrons. Chlorine (Cl) atoms all have 17 protons and 17 electrons, and this is what makes them chlorine atoms. But out of every four atoms of chlorine, three have 18 neutrons, or a mass of $18 + 17$, which is 35, and the fourth atom has 20 neutrons, making a mass of $20 + 17 = 37$. The two types of chlorine atoms are called isotopes of chlorine. Each one is an isotope. One is isotope Cl^{35} and the other is isotope Cl^{37}. The average mass of the two kinds is 35.453, which is its atomic weight. This ratio is worldwide, no matter in what compounds the chlorine is found—for example, salt. Chemically the two isotopes are identical, and physically they differ only in their mass. They are indistinguishable in ordinary chemical reactions.

There are three isotopes of hydrogen: the lightest has a proton only, another has a proton and a neutron, and the heaviest a proton and 2 neutrons. The ratio is 5000 of the first to 1 of the second and only a trace of the third, and their average atomic weight is 1.008. The heavier isotopes of hydrogen have been given names of their own because of their importance in nuclear research: deuterium and tritium. Deuterium's atomic mass is 2. Water molecules, one or both of whose hydrogen atoms are deuterium, are called heavy water. Tritium is radioactive and is made in nuclear reactors. Its atomic mass is 3.

Oxygen has three isotopes with atomic masses of 15, 16, and 17, but because oxygen was taken as the standard for atomic weights, against which all others were compared, the average atomic mass of these isotopes was given the round number 16 long before anything was known about isotopes.

Some elements have as many as seven isotopes. The only way isotopes of an element can be separated, a goal sometimes necessary in certain kinds of research, is by methods which distinguish among them by virtue of their different masses.

Sometimes an element has an isotope heavy enough to bring the average weight higher than the average weight of the next element, as is the case with cobalt (Co), atomic weight 58.93, and nickel (Ni), which follows it in the table although its atomic weight is only 58.71. When Mendeleev first lined up the elements, nickel necessarily came before cobalt, but their properties made him switch them around. There were several other such switches, which could not be understood until the reality of isotopes was accepted.

Isotopes simply do not come into ordinary chemical work because all the isotopes of an element are chemically alike. However, not only are there over three hundred known isotopes of the 92 elements, but many of these have been created in nuclear reactors of one kind or another and are radioactive. Their life span is often quite short; a few moments and they have disintegrated or transformed themselves into stable isotopes. It is these radioactive isotopes of certain elements that have become vitally important in biochemical research and they are becoming increasingly important in inorganic chemistry as well, in studies being made on the mechanism of what were once thought to be very simple chemical reactions, which are now considered to be very complex.

The usefulness of the periodic chart

The fantastic pattern that Mendeleev uncovered among the properties of the elements did not please minds that were most comfortable with absolutes, or perfect abstractions, or the formula precision that physics provides abundantly and chemistry to a lesser extent; biology and the other life sciences provided it not at all until recently, when the statistical view was sharpened into a mathematical tool. Anyone could see that the periodic table was full of exceptions, contradictions, vague generalities, and uneven trends, and that the pattern had to be picked out patiently against a screen of stubborn, independent facts. Today the first reaction to the table is the same: are these surprising resemblances worthy of serious attention? Are these obscure juxtapositions, counterpoints, runs, meaningful or are they imposed on reality by wishful thinking? The idea seems to strain after relationships that cannot possibly lead to laws.

On the other hand, the correlations were persistent enough to infect the endless speculations concerning the nature of the atom. Many attempts were made to produce a perfect table or at least a geometrically balanced arrangement, and even a three-dimensional cylinder was constructed, with the elements spiraling around it in a helix. Nothing could improve on Mendeleev's approach, for the simple reason that there was only one pattern and it rested on the atomic weights, which were not ideal numbers.

The ancient question of how many elements there might be also became fascinating again. At that time less than seventy were known, and all the rest have been discovered in the knowledge that the table provided, a knowledge which does not limit the possible number. It is an open-ended table.

A chart of resemblances is also a chart of dissimilarities, willynilly. The like and the unlike in nature and in thought are inseparable, and wherever we find the one, the other is integral. Every harmony depends on a particular selection from an orderly arrangement of apparently unlike notes, and this should be kept in mind whenever the periodic information that can be crammed into the chart is studied. It is all empirical and compiled from observation of phenomena that, though they do not yield a theory concerning the structure of atoms, allow us to calculate precisely, for example, what element 104 ought to be like.

The usefulness of the chart varied during its century of life. Most authors thirty years ago relegated it to an honorable place in history, others found it acceptable as a teaching aid, but no chemist needed it, for the simple reason that he knew far more in his own field than could be printed in any chart, and whatever he forgot or wanted to know about other areas he found more easily and in more up to date form in the handbooks of chemistry and physics. Then, recently, the table acquired new importance as an aid to thinking at the research level because of the proliferating work with radioactive isotopes, and because of the search for profitable exploitation of the rarer elements. In other words, the chart is not used in routine laboratory work but it is impossible to think without it, especially in research, when the probabilities of certain reactions are being explored. As for teaching and studying, neither can be imagined without the periodic table. Finally, all textbooks group the discussion of properties according to the patterns revealed by the table.

The relatedness of properties

Let us glance at a few of these patterns.

The boiling and melting points of any element or compound are figures so precise that they are used for absolute identification in many cases where other kinds of recognition are uncertain. In group VIIA, fluorine (F) boils at $-187°$ C. and thus is a gas at room temperature. Chlorine (Cl), almost twice as heavy as fluorine, is also a gas at room temperature but it boils at $-35°$ C. Bromine (Br), approximately four times as heavy as fluorine, is a liquid that boils at $+59°$ C. Iodine (I), six times as heavy, is a solid which vaporizes directly into a gas, that is, sublimates, at $+183°$ C. No mathematical relationship can be found to tie these boiling temperatures either to the atomic weights or to the number of protons. All that can be said is that as the atoms get heavier in this group, their boiling points rise. In other groups the trend may be reversed. Yet, in general, more often than not a heavy substance will tend to have a higher boiling point than a lighter element in the same group. This means that in general the boiling points at the bottom of the table will be higher than those at the top.

Similar observations can be made concerning the melting temperatures, hardness, crystal-structure type, color, viscosity, electrical conductivity, magnetic properties, the heat of fusion, and all other physical properties. But it is entertaining to note that smell and taste, which are of prime importance in identifying compounds in the laboratory, and which are always listed in the most severe catalogue of properties, cannot be reduced to numbers and thus cannot be categorized into a pattern.

The most surprising pattern is that for the diameter of the atoms. They do not get steadily larger, as we shall see, but the pattern, again, could not have been reasoned out before it was obtained by measurements. It explains a great deal about the behavior of atoms in chemical reactions. In fact, the discovery of the effect that the relative size of atoms has on their chemical and physical natures recently has reduced the difference in conception between these two categories to an extent that would have been inconceivable thirty years ago. We can actually predict some kinds of chemical activity from the relative diameters of atoms. However, it will always be useful to distinguish between physical and chemical properties of bulk matter, simply because they are distinctions that our unaided senses can make.

But let us consider some fresh aspects of atomic structure, using the periodic table which can be imagined as a chest of drawers, each drawer labeled with the symbol for an element, and in each drawer all the isotopes of that element.

We know that when a shell (except for the K shell) has acquired 8 electrons a new one begins outside it, whether or not it is filled to capacity, and thus the shells M, N, O, P complete themselves inside a partially filled outermost shell. This means that many elements will have the same number of *outermost* electrons and ought to be in the same vertical group. But other considerations make some of the electrons inside the outermost shell significant, so that such elements can be distributed into the eight groups as though they really had 3, or 4, or 5, and so on, outer electrons, even though they may have only 2. However, in the very longest period, the sixth, when shells O and P are being filled inside, a set of fifteen elements turns up that are almost identical, and they all have to be put into one box in the table, even though each one has a different number of protons. They were named the rare-earth elements because of their presumed scarcity, but today they are called the lanthanide series, after lanthanum, the 57th element, which is the first in the set and which thus represents them all. To stretch the table to accommodate them would be illogical, since the element that fits into the next group is hafnium (Hf), element 72. Therefore the fifteen have been plucked out of the table and listed separately below it. The same procedure has been followed with the fifteen elements of the so-called actinide (Ac) series in the seventh period, comprising the heaviest elements, from actinium, the 89th element, to uranium (U), the 92nd, and all the new man-made elements following as well.

Element 101 was at last named mendelevium.

PART III

The Chemical Nature of Atoms

* *

13

Electron Shift Is Chemical Change

The most and least active atoms

ONE of the most important properties of an element concerns the general exuberance with which it reacts with other elements of its own and other groups. There are several ways of defining this exuberance, or willingness, or eagerness to be active, or to enter reactions—to combine. The converse of great vitality and passion for reacting would be the idea of a placid, or stable, or unexcitable, or inactive nature, inert and preferring solitude, something resistant to change, not tempted or pushed into combinations. Using this idea, the degree of activity within a vertical group varies in a fairly regular manner, and the over-all pattern of activity of any one group differs measurably in its specific aspects from the over-all pattern of activity of the groups next to it.

A certain amount of close study is necessary before the essential points of the periodic table can be retained by the memory, but for the moment only the following aspect need be kept in mind.

In each group there are both A and B elements. Originally these were put in the same box, as in the modernized version of the Mendeleev table on pages 114 and 115, but as atomic structure became clearer the long form we see in the table on pages 100 and 101 came into use. In it, A and B subgroups have been separated. As far as most of our discussions in this book are concerned, we need to see clearly only the groups IA, IIA, IIIA, IVA, VA, VIA, VIIA, and the inert gases, which are group O. Let the reader fasten his gaze therefore on the two end groups at the left and the six end groups at the right, and let him assume that all the groups in between will fit into everything that is said about these end groups.

The elements of groups IA and VIIA are the most active of all the elements, meaning that they will participate in a great many reactions with enthusiasm and with speed and will produce many and varied and stable compounds which are also very active.

The elements that lie between these end groups in the table are considerably less active or spontaneous in their behavior, and on the whole they form fewer compounds, with the exception— there are always exceptions in this table—of carbon (C) in group IVA and of oxygen (O) in group VIA. Carbon is such an exception that a study of all its compounds has become the special science of organic chemistry. Nevertheless, since the proliferation of carbon compounds depends upon a special bond between carbon atoms, and since carbon does not form many compounds with many other elements, carbon is a much less reactive element than chlorine, for instance.

If there are most active groups, there must be a least active group, and there is one, group O, starting with helium. We have already mentioned the nature and the names of these elements—the rare gases, the inert gases, the noble gases. All of them are so inactive that they form not a single compound under ordinary laboratory conditions and they remain aloof even from themselves, being atomic in the gaseous state, something no other gas can be. Recently some compounds of the rare gases have been created by means of extremely high pressure, and thus the assertion that they are totally inactive, which is what was believed until now, has to be modified. Obviously their inertness remains striking, and not another element among the 92 comes even close to their stability. They even refuse to liquefy except at extraordinarily low temperatures.

If we now assume that the properties of an atom are largely the result of its electron structure, then we must conclude that having 8 electrons outermost seems to produce an extremely stable atom. That outer shell of 8, which all the rare gases have, cannot be broken even by increasing the number of protons in the nucleus; a new shell has to start outside it until the larger nucleus is strong enough to pull the new electrons down. To refresh the reader's memory: it is a question of overlapping shells, of the lowest available energy in the *s, p, d, f, g* subshells, that determines where the new electrons go.

We can say that in the inert gases the energy distribution is so perfectly balanced that almost nothing chemical can disturb it. An atom with 8 electrons outermost is almost free of inner tensions, as it were. There is no loose, restless, excess, free available energy inside such an atom. Therefore this group is given the number O, to indicate its passive nature. It is the key to the periodicity of electron shells and to the resulting periodicity of the properties of the elements. Once a system has 8 electrons outermost, it has less available energy than at any other time. It is as locked up, in a sense, as it can ever be.

We have not discussed energy in mathematical detail but we have mentioned it enough times so that perhaps it has become a little familiar as a concept associated with movement and heat and radiation, and with magnetism and electricity. We have also stated quite simply that twentieth-century science was born when it was discovered that energy is not a continuous, steady flow, but consists of finite quanta in the same way that matter consists of atoms. A piece of matter emitting energy in the form of light will be emitting "particles," or quanta, called photons of light. A piece of ice in my hand will melt because energy in the form of heat—molecular motion—will flow from, or be transferred from my hand to the ice. The ice has heat in it too, but, according to a law of thermodynamics, that heat will never flow into my hand, which contains more heat. A billiard ball hitting one that is moving more slowly will speed it up and will be slowed in turn by an amount equal to the energy it transmitted. A slow ball will never speed up a fast ball.

Upon examination of the phenomenon of transfer of energy, we have to conclude that, no matter what kind of energy we are talking about, it will be transferred from a system that has more to a system that has less. Always. Of course, energy can be forced from one system into another by proper application of an outside force. But a system with a lot of energy will never take into itself, of its own accord, energy from a system that has less. In the absence of an applied force, energy will not flow from a small reservoir to a larger one. The piece of ice will always cool my hand and it will always be heated by my hand. (We may note that modern science-philosophy has introduced a measure of philosophical doubt into this concept.)

The concept of entropy

The idea of a one-way flow of energy has been incorporated into an abstraction called entropy, and some remarkable conclusions have resulted. If every system gives up its energy only to systems of lower energy, then eventually everything will arrive at the same common-denominator level of energy. An undifferentiated uniformity will prevail throughout the universe. Every particle will have exactly the same energy as every other particle. This uniformity is called total chaos, for within it there cannot be discerned any organization, any order of any kind. The ultimate end of the universe will thus be a uniform distribution of uniformly endowed final particles mutually incapable of communicating anything, since not one can either take or give energy to another. This terrifying vision is logical, and entropy lends itself to some other formulations that are also pitiless to the human spirit. There are forces in the universe which can be interpreted in such a way that they modify the concept of entropy, and the concept does not eliminate the possibility that, for a time, locally ordered systems may arise; for instance, a living organism does enlarge its total energy by an orderly growth process; but the over-all view at present is that if the universe is not turning into chaos, then some still undiscovered force is at work preventing it. Insofar as science is concerned, the statement that systems of high energy content always give up energy to systems of low energy content holds with a truthfulness that has no exception.

Without exploring this fascinating idea further, we turn back to the system of energies within the atom, and specifically to the energies associated with the electron cloud.

We must not forget that every atom above $-273°$ C. is also imbued with motion that the kinetic molecular theory defines and that this kinetic energy of the total atom is not what we are discussing at the moment. The kinetic energy of an inert-gas atom is always available. The inert-gas structure is stable not because it has no energy in it but because the various parts are so harmoniously fitted together that the whole atom has all the available energies within it meshed together without strain.

Consider an element just before any inert gas in the periodic table, and an element just after any inert gas. Fluorine (F) in group VIIA precedes neon (Ne), which is followed by sodium (Na) in group IA of the next period. Their atomic numbers are 9, 10, and 11 respectively.

We have said that both fluorine and sodium are extremely active elements. They belong to the most active groups in the table and they combine with many other elements and compounds vigorously, liberating a great deal of energy in the process. In general, one can say that an atom is active because it is loaded with excess energy that it can offer or make available to a reaction. Fluorine and sodium are stable atoms, but they are systems with a lot of energy they can get rid of easily when another system having less will accept it. After the exchange, fluorine and sodium will have less "free" energy and the receiver will have more. Between these two structures eager to rearrange themselves in order to shed some of their energy is the structure of neon, which resists all ordinary attempts to make it yield energy.

Ions

If we now imagine removing the single electron in sodium's outermost shell, the M shell, the configuration of the remaining electrons will be identical with that of neon. Having removed that electron leaves only 10 in sodium, which has 11 protons. Thus one proton is left unneutralized, and the particle as a whole has one excess positive charge on it. Can such a particle exist? Indeed it can and, in fact, sodium is very eager to get rid of that extra electron and with it some of its energy, and thus to acquire the more stable configuration of neon. With its outermost electron gone and a positive charge on the whole provided by the unbalanced nucleus, this residue of the neutral sodium atom is called an ion, an ion with a positive charge, or a cation.

The sodium atom's electron can be removed by a variety of methods. In a strong electrical field the electron will leave the atom and move to the positive pole of the field. At very high temperatures the electron will become excited enough to leap out of all its permissible orbitals around the nucleus and depart forever, a freely zooming electron, taking away with it some of the total energy of the atom. If the sodium atom is jarred violently by collision with another particle, even by a freely zooming electron, that solitary outermost electron can be sheered off. Certain kinds of radiation will act partly like a solid blow, and the electron will become so excited that the nucleus will not be able to hold it. In every case, as the electron departs, a certain amount of energy is released by the atom as it rearranges its inner energies into a more stable mesh.

But to us the most interesting and by far the most common way in which sodium will lose its electron and become a positively charged ion is through ordinary chemical reaction. That is, the nucleus of another atom will simply remove the electron from sodium's cloud and attach it to its own.

Now consider fluorine, the element with 7 electrons in its outermost shell, the L shell. It lacks only one to complete this shell and acquire the same neon configuration of electrons that sodium acquired by losing an electron. Instead of yielding any of its electrons, fluorine seeks for one to put into its outermost shell. There are electrons available in several situations. An electrical field, for instance, that of a battery, will pile up electrons at the negative pole and a fluorine atom can seize one there easily. When it does so, it will release a certain amount of its inner energy. It is still fluorine because the number of its protons has not changed: it has 9 protons, but it now has 10 electrons. The particle therefore has an over-all negative charge. It has become an ion, a negatively charged ion, or an anion. This negative ion of fluorine has lost some of the neutral atom's energy and is therefore less active, just as the sodium ion which is positive is less active than the sodium atom was.

We can put these events into some kind of shorthand quite easily.

"Sodium atom loses one electron to become negative sodium ion" can be written:

$$Na \text{ minus } 1e^- \rightarrow Na^+,$$

and this can be rearranged to:

$$Na^0 \rightarrow Na^+ + e^-.$$

In the same way we can symbolize what happens to the fluorine atom:

$$F^0 + e^- \rightarrow F^-.$$

Fluorine can acquire its electron in a chemical reaction from any element willing to give it one. Such an element, one that yields electrons, is sodium. Fluorine and sodium, therefore, ought to react to each other's needs. The ions of both these elements are more stable than the original atoms were; that is, the ions will not alter themselves as eagerly as the atoms did. But before we go deeper into such a possible event, let us examine a few other elements.

Any member of group IIA, we know, has 2 electrons outside an inert-gas configuration. Can both of these be removed? Yes, and not only can both be removed but if the first one is removed, the second inevitably is. The structure of magnesium (Mg) cannot be disturbed without its losing both outermost electrons to become a cation with two positive charges. In group IIIA the elements lose all three of their outermost electrons and the atom becomes a cation with three positive charges.

All group VIIA elements need one electron to complete their eight, or octet. Group VIA elements—for example, oxygen—have six electrons outermost. Such an atom will not add just one electron to its outer shell; if it accepts any it will take two, to arrive at the octet. It will then be an anion with two negative charges. All the elements of group VA, having five electrons, will accept three electrons to count eight in their outer shell, and they will become anions with three negative charges.

Group IVA elements have a choice: to lose four or to gain four electrons. Actually they will neither gain nor lose in the manner that we have just described. In a sufficiently strong field their four electrons can be torn off, but in ordinary chemical reactions this cannot happen.

Once again there is a lovely simplification here that, also once again, will gradually become qualified in many ways. For the moment we will accept the simplification as the whole story.

Transfer of electrons

The word used to describe these electrons which an atom can lose from its neutral self, or add to its neutral self, the electrons that make the difference between the atom's structure and the structure of the nearest inert gas, is valence. The valence electrons are those that an atom can most easily gain or lose, and the valence shell is the one in which these movable electrons are.

We have already met an earlier concept of valence when the word was coined, long before anything was known of the electron structure of the atom, to define the combining power of an element. Hydrogen was described as having a combining power of 1, while oxygen had a combining power of 2. This meant that each oxygen atom could clasp two hydrogen atoms. Every element had a combining power or valence of 1, or 2, or 3, or 4. The concept was symbolized by giving each atom as many hooks as it had valences. In forming molecules, all the hooks of the atoms involved had to be engaged.

This view could not explain many of the simplest compounds and none of the complex ones, yet it was so useful that it was impossible to discuss chemical reaction without valence. When the electron structures were worked out the word was kept and changed to mean, literally, the number of electrons an atom will lose or gain in a chemical reaction.

We now have two versions of every element, the neutral atom and its electrically charged ion, which the atom forms by losing or gaining electrons. Are the properties of an ion much different from those of its atom? Yes, emphatically yes. The two particles are altogether different in their properties.

Let us imagine a sodium (Na) atom and a chlorine (Cl) atom colliding. The sodium atom, though enduring if left alone, and perfectly stable if not in contact with other atoms, nevertheless eagerly gives up its single valence electron in the M shell to whatever is willing to accept it. The chlorine atom is also a very solid piece of enduring matter that will roam about forever without changing if it meets nothing that will provide it with an electron. But it is highly tense and "eager" to reduce its energy by acquiring a valence electron to complete its shell.

If chlorine gas and sodium metal are in contact and their atoms are sufficiently excited, the sodium atom will get rid of its single valence electron by donating it to chlorine, which is eager to add one electron to its outer shell. The two resulting particles will be a sodium cation and a chlorine anion, which, because of their opposite charges, will cling together and produce the compound sodium chloride, or ordinary table salt.

The two poisonous elements have reacted chemically to form a compound with extraordinarily different properties, merely by transferring a single electron from one to the other.

A vast number of chemical reactions consist of nothing more than an actual transfer of valence electrons from the electron cloud of one nucleus to the electron cloud around a different kind of nucleus.

Suppose we take calcium (Ca) and oxygen (O). The reaction when two of these atoms collide consists of a transfer of two electrons from calcium to oxygen. The oppositely charged ions of calcium and oxygen, which can be represented as Ca^{++} and O^{--}, will cling together and form a compound, consisting of molecules in the gaseous state, and a latticework of positive and negative ions in the solid state.

To visualize a reaction between chlorine and aluminum requires a little arithmetic, since aluminum has three electrons to give, while chlorine can accept only one. What happens is that aluminum gives one electron to each of three atoms of chlorine.

If we bring aluminum and oxygen together, arithmetic reveals that two atoms of aluminum with a total of six valence electrons will combine with three atoms of oxygen, which together will accept six electrons. The compound will consist of ions in the ratio of 2 Al and 3 O, and we write this Al_2O_3.

In all such reactions the transformation of properties that we outlined for chlorine and sodium are paralleled. The properties of the atoms involved are wholly different from the properties of the ions and also from the properties of the compound that the ions form. The properties of the reacting elements vanish, and a totally new substance appears with its own specific properties, which belong exclusively to the linked ions. The magnetism, hardness, luster, density, the "ironness" of iron, as it were, fuses with the yellow, powdery, inflammable, stink-producing "sulfurousness" of sulfur to produce a black, granular matter that has no magnetism, no ironness of any kind, no sulfurous quality at all. The invisible movement of a few invisible electrons from one invisible atom to another causes the transformation.

This particular kind of chemical bonding is called *electrovalent*, or *ionic*. Electrovalent, or ionic, bonds are one of the three major binding mechanisms in chemical reactions.

Metals and nonmetals

Before we proceed to the second type of bond, it is useful to divide the elements into two large categories. The elements that lose valence electrons in order to uncover inert-gas configurations, because they follow an inert gas in the table, are called metals. The metals are therefore found on the left-hand side of the table —groups IA, IIA, and IIIA. The elements that gain electrons to complete their valence shells so that they are like the inert gases that follow them are called, with great brilliance and ingenuity, nonmetals. The nonmetals are the elements in groups IVA, VA, VIA, VIIA, and O. The B groups will be mentioned later.

The two kinds of elements are distinguishable in many ways, but the source of their difference lies in the fact that metals will give up electrons, while nonmetals will accept them.

We mention here, and explore more deeply later on, the fact that simply a collision between a metal and a nonmetal does not necessarily produce a reaction. The two atoms have to be prepared to give up and to accept electrons, and there are definite limits to the conditions under which a particular reaction will take place. Usually the limit in the laboratory is temperature. For each reaction there is a temperature below which it will not take place. Often there is a temperature above which the desired reaction cannot win out, but all this has to be examined later. Stove gas illustrates the usual phenomenon. It will not burn until its temperature is raised by the flame of a match.

Sharing of electrons

The second manner in which atoms combine is called *covalent bonding*. The nucleus of each of the elements exerts a binding force on its electrons, which increases as each electron is removed. Normally the unbalanced positive charges on any nucleus cannot build up to more than three. It is extremely difficult to pull off the fourth electron in any valence shell, against the positive pull of four unbalanced protons. This is why any element with four, five, six, or seven electrons must always add electrons to its valence shell instead of stripping them off.

Yet we find that nonmetal atoms combine with one another into molecules that are much more rigidly bound than are ions in electrovalent compounds. Since the atoms of nonmetals can only seize electrons, the only explanation is that they seize but must share each other's without actually transferring them.

What happens when two fluorine (F) atoms collide? Each atom needs one electron to complete its valence shell. One electron from each atom swings out to revolve around the other atom, too, in its gyrations. These two valence electrons literally sew up the two atoms. Each atom now counts eight electrons in its cloud, even though two of those eight spend some of their time around the partner.

This kind of chemical bonding is called "covalent" for obvious reasons. Covalent bonds do not produce ions, but the transformation of the properties is just as thorough. Atoms of fluorine are quite different from molecules of fluorine. The atoms are so active that if they are liberated in some chemical reaction they instantly seize whatever electrons are handy—each other's, if they are the only available ones, to form covalently bonded fluorine molecules, which can be stored indefinitely.

Another familiar example of covalent bonding occurs when coal, wood, oil, fats, or any organic fuel burns, breaking down to combine with oxygen. The products are carbon dioxide and water molecules, both covalently bonded, neither of which resembles even remotely the oxygen and the covalently bonded fuel molecules that reacted with each other.

If we bring together two hydrogen atoms and one oxygen atom, and let them combine into a molecule of water, what has happened? A hydrogen atom could gain one electron to complete its K shell and become a negative ion; it could lose its single electron to become a positive ion, or, in effect, a solitary proton; it could share its electron with an atom in return for the loan of an electron from the other atom, thereby creating a covalent bond. Hydrogen can thus be placed either above group IA or above group VIIA. It can act as a metal or as a nonmetal, depending on the conditions. With oxygen, hydrogen forms covalent bonds. With strong metals hydrogen forms ionic compounds, for instance $Na^+ H^-$. Theory alone could not have told us, but research brings in irrefutable evidence.

There is a kind of covalent bonding in which one of the reacting atoms offers to share two of its electrons with another atom that is willing to accept the offer of both. The resulting link is to all intents and purposes a covalent one, with the pair of electrons being shared by two nuclei. To distinguish the origin of the shared pair, this kind of bonding is called coordinate covalence. The bonds in a great many molecules could not be unraveled without the recognition of coordinate covalent links. In this book the difference between coordinate covalence and covalence is not emphasized.

The third major method of bonding is called *metallic bonding*, which is discussed in a later chapter. But we can now make a few summaries concerning chemical reactions in general.

All metals and nonmetals that combine do so by transferring electrons from metal to nonmetals, thus forming compounds of ions which cling together in electrovalent bonds.

All nonmetals that combine with one another do so by sharing one or more pairs of electrons to produce covalently bonded molecules. Covalent bonds are possible only between atoms that have four or more electrons in their valence shell. Elements with less than four electrons cannot form covalent bonds.

Nonmetals have a negative valence number, indicating the charge on the anion, while metals have a positive valence number, indicating the charge on the cation. Nonmetals also have a co-valent number, indicating the number of electrons they will share.

All elements of groups IA and IIA are strongly metallic; elements of groups VIA, VIIA, and O are strongly nonmetallic. The light elements of groups IIIA, IVA, and VA are nonmetals but the heavier ones in these groups are called metalloids because they behave like nonmetals only under certain conditions, the rest of the time their properties being metallic. Thus, no sharp borderline exists in the table between strong metals and strong nonmetals.

By far the largest number of elements belong to the B groups and group VIII and these are called the transition metals. Their behavior with respect to valence electrons is complicated by the fact that they have incompleted inner shells building up within an outermost s orbital. That is, their outermost shells never contain more than two electrons, sometimes only one, but they are able to yield more than these, from inner d and f subshells. Thus the number of the B group into which they fall by virtue of their atomic number does not indicate their valence, which is generally $+1$ or $+2$ or $+3$. Furthermore, while succeeding metals and non-metals in a period differ a great deal in their properties, succeeding transition metals in a period are often much alike. For example, most of them are paramagnetic, that is, they are affected by magnetic fields, a property related to the unmatched spin of electrons in the inner d and f subshells. All the metals which are familiar to us in our daily experience are transition metals or metalloids.

At this point we pause in our delving into the atom and enlarge our view to examine the symbolic language of chemistry, without which nothing would be comprehensible to the chemist. Symbols, formulas, and equations seem to be the very essence of chemistry.

14

The Alphabet and Grammar
of Chemical Information

Symbols, formulas, and equations

THE law of conservation of matter states that all the particles that enter a reaction will weigh exactly the same as all the particles that are the products of that reaction; the law of conservation of energy states that the total energy content of the reactants and of all other energies used in a reaction will equal the total energy associated with the products after the reaction stops. These laws were enunciated some time after Lavoisier established gravimetric procedure as the only ground on which chemical reactions could be understood, and most recently in nuclear physics it was proved that they have to be stated together: the sum total of mass and energy at the beginning of a reaction equals the sum total of mass and energy when the reaction ends, or at any point where it is stopped. Hence, a chemical reaction cannot destroy or create matter and energy, and it is on this supposition that symbols for the elements and other chemical notations are valid.

The symbol for an element consists of a single capital letter or of a capital and one other letter, derived from the name of the element. Since many elements were known in antiquity, their symbols reflect those ancient names, for example, Au for gold comes from the Latin *aurum*. Elements discovered in the last few centuries were usually given Greek or Latin names signifying some property, and most recently the names of scientists have been used, with Latinized endings.

The symbol for an element represents a single neutral atom of it. By this we mean not any hypothetical essence of the element but a real atom with all its protons and neutrons and electrons, all its mass, and all its energies. Chemical symbols cannot be used in any other way. They are statements of quantity. For instance, if we say, "5 grams of Na" we mean 5 grams of sodium atoms. But statements such as "cut up a piece of Na" are meaningless. The symbol cannot replace the name of the element as a kind of shorthand device. We can say, "Na and Cl react," because that statement is true; sodium and chlorine atoms do react. To indicate more than one atom, the number of atoms is put before the symbol: 2 Na, or 5 Cl. The symbol also represents the atomic weight of the element.

A formula in chemistry is the symbolization of a molecule or part of a molecule. Chlorine atoms combine covalently to form molecules with two atoms each, and the formula for a single chlorine molecule is Cl_2. This indicates not just the idea of one chlorine molecule composed of two chlorine atoms; it actually stands for the total mass of such a molecule. The atomic weight of chlorine is 35.453, thus the molecular weight of chlorine is 70.906. The number of atoms present in a molecule is always indicated by a small number to the right of the symbol and dropped below the line. The formula for sulfuric acid is H_2SO_4, which means one molecule of sulfuric acid consisting of two atoms of hydrogen, one atom of sulfur, and four atoms of oxygen, all in covalent bonds. The molecular weight of this compound is the sum of the atomic weights: hydrogen 1.008×2 + sulfur 32.064 $\times 1$ + oxygen 16×4, all of which equals 98.08. If we want to say three molecules of sulfuric acid, we write $3 H_2SO_4$.

Suppose we have a compound whose molecular composition we don't know. By appropriate analytical procedures, the elements can be broken out of their bonds in the compound and their ratios determined by weighings. For example, when carbon dioxide,

CO_2, is analyzed we find that 27 per cent is always carbon and 73 per cent always oxygen. Using the relative weights and a little arithmetic, we calculate that there must be a ratio of 1 atom of carbon to 2 atoms of oxygen in the molecule. The actual molecule might contain 5 C and 10 O, but proper research determines that the formula is CO_2.

Symbols allow us to indicate, by superscript, the particular isotope we may be working with, thus U^{235} or H^2 or O^{17}. If there is no such superscript then we know that the symbol stands for the world average of that element's isotopes as reported in the handbook of chemistry and physics.

The symbol can also be used to indicate the atom's ionic state, by adding a small plus or minus sign as a superscript. The positive ion of sodium is Na^+ and the negative ion of chlorine is Cl^-, of oxygen O^{--}. The cation of aluminum is Al^{+++}. Thus the compound sodium chloride can be represented by the formula Na^+Cl^-.

Now let us turn to chemical equations in which symbols and formulas are used.

We know, to use what we have already discussed, that one atom of chlorine and one atom of chlorine will react by sharing an electron each and will become a molecule of chlorine; and we know that the actual number of protons, electrons, neutrons before the reaction must equal the number after. Therefore we can write:

The mass of 1 chlorine atom plus the mass of 1 chlorine atom equals the mass of 1 chlorine molecule, or

$$1\, Cl + 1\, Cl = 1\, Cl_2.$$

But we also want to show that this is not just a static statement of sums but that a reaction has taken place. We substitute for the equation sign an arrow which has two meanings: it is an equating sign that also indicates that a reaction can take place in the direction the arrow points. We can drop the figure 1, since each symbol by definition represents one atom or molecule:

$$Cl + Cl \rightarrow Cl_2.$$

The plus sign now has its meaning enlarged, too. Not only does it refer to an addition of weights but it also indicates that the two atoms have actually participated in a fruitful collision. Every equation stands for a reaction that has been proved to take place. The equation for a hypothetical reaction, of which there are obviously as many as we care to invent out of pure fantasy, must be clearly labeled hypothetical. We cannot ever swear that we can predict absolutely the course of a new reaction and, until it has been verified, no equation can be written for it.

The chemical equation, in other words, is a clear and rigid mathematical statement of actual brute facts, and its substance is never in any way meant to be a descriptive analogy. It is absolutely quantitative and is as precise as any law that can be formulated in mathematical symbols.

Balancing the budget

Consider another aspect of reactions we have already discussed. One atom of sodium combines with one atom of chlorine to form one ion of sodium and one ion of chlorine, which cling together in electrovalent bonds to form salt, which might be expressed:

$$Na + Cl \rightarrow Na^+ + Cl^- \rightarrow Na^+Cl^-.$$

In reality, since atoms of chlorine are too active to exist free, we have available only diatomic molecules of chlorine gas. This seems to indicate that the reaction should be written:

$$Na + Cl_2 \rightarrow Na^+Cl^- \text{ (not balanced)}.$$

But something is obviously wrong here, because it violates our definition of an equation: the number of atoms on the left does not equal the number on the right. With a little bit of arithmetical reasoning we multiply every symbol in the basic equation by 2, and write:

$$2Na + Cl_2 \rightarrow 2\,Na^+Cl^- \text{ (balanced)}.$$

The equation is now balanced.

When coal, which if it is of good quality is mostly carbon, is burned in air to produce heat, the following reaction takes place with oxygen:

$$C + O_2 \rightarrow CO_2 \text{ (balanced)}.$$

If the oxygen supply is reduced to half of this, we get carbon monoxide instead:

$$2C + O_2 \rightarrow 2CO \text{ (balanced)}.$$

A little study of the arithmetic in these two reactions will help the reader see what is meant by a balanced equation.

Suppose we want to know how much oxygen is needed for the first of these reactions when one ton of carbon is burned. The atomic weight of carbon is, in round numbers, 12, of oxygen 16. The molecular weight of oxygen is therefore 32. Wherever a symbol stands we can substitute for it the atomic-weight ratios:

$$C + O_2 \rightarrow CO_2$$
$$12 \quad 32 \quad\quad 44.$$

The weight ratio in which C and O_2 combine is 12 to 32. We can use any system of weights we choose in this ratio, which stands for the relative weights of the atoms involved. For instance, 12 grams of carbon use up 32 grams of oxygen, to produce 44 grams of carbon dioxide. Or, 12 tons of carbon need 32 tons of oxygen to burn completely. By division we calculate that 1 ton of carbon needs $2\frac{2}{3}$ tons of oxygen, or about 11 tons of air.

Two atoms of hydrogen combine with one atom of oxygen to form one molecule of water. Since in reality both hydrogen and oxygen are diatomic molecules, the balanced equation is:

$$2H_2 + O_2 \rightarrow 2H_2O$$
$$2 \times 2 + (16 + 16) = 2 \times (2 + 16).$$
$$4 \quad\quad 32 \quad\quad\quad 36$$

Hence, 4 ounces, or 4 pounds, or 4 tons of hydrogen use up 32 ounces, or 32 pounds, or 32 tons, respectively, of oxygen when they combine. Or, 1 unit of hydrogen—say, 1 kilogram—needs 8 kilograms of oxygen; or, 20 pounds of hydrogen use 160 pounds of oxygen.

These few examples suggest the most important use of equations, and the point to remember is that every electron, and every nucleus, must be rigidly accounted for. This was the gravimetric point of view that Lavoisier introduced with such monumental certainty, a generation before Dalton's atomic theory, and it is the ground on which all chemical work is based, whether in research or in industry. All knowledge of a chemical nature is obtained by the use of equations, and it is from equations that the vast installations are built that manufacture pharmaceuticals, synthetic fibers and plastics, foods, metals, cement, gasoline and petroleum products, alcohol, acids and caustics, just to name a few of the numberless substances that are absolutely vital to our way of life. It all goes back to the concept that a chemical reaction can be represented perfectly as a mathematical accounting of weights of the elements and compounds involved. The practical application of equations to measurement is called stoichiometry.

The use of the chemical equation is proof of our absolute faith in the orderliness of chemical transformations, and in their complete predictability once they have been charted. Without the chemical equation any concept of change in the properties of matter is as fruitless as it was for the alchemists.

What about the energies involved? In many kinds of industrial processes the energy produced by a reaction is the sole reason why it is useful, for example, the burning of fuels for heat. We know that when carbon burns it produces flame and heat.

$$C + O_2 \rightarrow CO_2 + E_{heat}.$$

This does not mean there was no energy associated with the C and O_2 on the left-hand side of the equation. That notation E_{heat} stands for a specific quantity of energy that was liberated by the reaction from the carbon and oxygen atoms. The molecule of carbon dioxide contains less energy locked up in it than the carbon and oxygen atoms had together, and this difference can be entered in the equation whenever energy factors interest us. It must be emphasized that the E in the equation is expressed in some unit of measure, such as calories. It is as specific and as precise as the weights used. Usually when large quantities of coal or wood or oil are burned for heat, the energy is expressed in British thermal units, which can be converted into kilowatt hours or horsepower, since energy is a force that does work.

We must point out again that no formula can be arrived at theoretically. Without exception, formulas are empirically derived, that is, from hard information. In the same way, all equations are empirical. It is impossible to anticipate what will happen in a reaction that has never been studied. By definition, until an unknown reaction has been made to occur and its details have been observed, no equation can be written for it. Therefore, it is impossible to predict theoretically the energy changes that will go on in an unknown reaction.

Of course our knowledge is so oceanic that a new compound's formula can be anticipated to a certain extent, and a new reaction's course can be guessed at. Almost no research is a wild, unreasoned shot in the dark, and a great many of the unpredictables are today probabilities. Nevertheless, formulas and equations are quantitative statements derived exclusively from observation and measurement.

Reversible reaction

We now introduce a concept that will require expansion later. Can a reaction be reversed? The answer is yes. In the reaction

$$2Na + Cl_2 \rightarrow 2\,Na^+Cl^- + E_{heat}$$

sodium metal and chlorine gas react, with the liberation of heat, to produce salt. If the salt is melted and energy in the form of an electric current, which is a source of free electrons, is passed through it, the sodium metal and the chlorine gas are regenerated from the salt. We can therefore write:

$$2\,Na^+Cl^- + E_{electrical} \rightarrow 2\,Na + Cl_2.$$

When shining mercury (Hg) combines with oxygen it forms mercuric oxide, a reddish powder, according to the equation:

$$2\,Hg + O_2 \rightarrow 2\,Hg^{++}O^{--} + E_{heat}.$$

The mercury atom transfers its two valence electrons to the oxygen atom, and the two oppositely charged ions then cling together. But suppose we now take the red powder and heat it to a higher temperature? It breaks down, yielding mercury and oxygen.

$$2\,Hg^{++}O^{--} + E_{heat} \rightarrow 2Hg + O_2.$$

We can therefore write this equation with a double arrow to indicate that it can go in either direction, depending on the amount of heat energy present.

$$2\,Hg + O_2 \rightleftarrows 2\,Hg^{++}O^{--} + E_{heat}.$$

This is the first experiment that a student is given in a beginner's laboratory. It is a very pretty sight when the reddish powder in a test tube, heated over a Bunsen Burner flame, gradually alters its appearance and coats the glass sides of the tube with silvery metal, while the oxygen coming out of the tube makes a smoldering splinter burst into flame.

This was also the experiment that Lavoisier studied to discover the correct answer to the nature of burning.

A reaction that releases energy—and it is almost always in the form of heat—is called exothermic. A reaction that absorbs energy is called endothermic. Most useful chemical reactions are exothermic, for the simple reason that they will proceed on their own steam once they have begun, while an endothermic reaction must be supplied constantly with energy to keep it going. But this is only a preliminary glance at the energy changes which are the key to all chemical reactions. Every shift in the configuration of the electron cloud around a nucleus is brought about by the atom's seeking or taking advantage of a condition in which it can reduce the total amount of its energy. Can all reactions be reversed merely by supplying enough energy, as in the illustrations we have shown? No. We cannot yet reverse the burning of natural fuels, for instance, or the digestive processes in living organisms. On the whole, the majority of the reactions which are industrially useful to us can be reversed by judicious manipulation of the conditions, but, as we shall see, the reversibility of most reactions is a problem, rather than an asset, in industry.

Before we narrow our view back to the electron clouds, a few more aspects of equations should be mentioned. In many reactions between two liquids a solid is produced which is insoluble in the liquids and which settles out. This can be indicated in the equation:

$$AgNO_3 + NaCl \rightarrow AgCl\downarrow + NaNO_3.$$

The silver chloride, $AgCl$, precipitates as a white crystalline solid.

If a gas is produced, this can be indicated with an arrow pointing in the opposite direction:

$$2HgO \rightarrow 2Hg + O_2\uparrow.$$

A reaction which must be heated to keep it moving, and in which the actual amount of heat provided does not interest us, can be written:

$$2HgO \xrightarrow{\Delta} 2Hg + O_2\uparrow.$$

Solids can be underlined $\underline{Na_2SO_4}$, and gases can be written $\overline{CH_4}$.

Equations, like numbers, can be added. We can write partial equations for what happens in a reaction and then piece them together. For example, sodium loses its valence electron:

$$Na \rightarrow Na^+ + e^-,$$

while chlorine gains a valence electron:

$$Cl + e^- \rightarrow Cl^-.$$

If we add the two:

$$Na + Cl + e^- \rightarrow Na^+ + Cl^- + e^-,$$

the electrons on both sides of the arrow cancel each other out, just as in any algebraic equation.

What we must always remember about equations is that electron configurations around the nuclei listed on one side of the arrow have become rearranged on the other side around the same nuclei, and in the process have shifted the electronic or covalent bonds between the atoms. Under one set of conditions the electrons will shape mercury and oxygen, and under another set of conditions they will shape mercuric oxide. All three can exist together in the same test tube. If we mix the powder and the mercury in contact with air—nothing happens. If the gas is turned on in the oven its hiss can be heard and it can be smelled. The oven fills up with gas; then the kitchen fills up with gas. Presently the whole house is full of gas, each molecule of gas smashing into molecules of oxygen millions of times a second. Nothing happens. But light a match and the gas begins to combine with the oxygen at explosive speed, with the evolution of enormous quantities of heat.

It is this sequence that we shall focus on, after a short digression.

What's in a name?

This is an appropriate place to comment a little on the nomenclature of inorganic compounds. Let us turn back to the periodic table. The general rule for naming any inorganic compound is to line up the elements that compose it, with the strongest metal first and the strongest nonmetal last. With compounds that contain only two kinds of elements, called binary, the ordinary name of the more metallic element is given first, then the name of the less metallic one is given, with the ending changed to -ide. The only exception is oxygen, which is always last.

Sodium chloride is NaCl, calcium bromide, $CaBr_2$, magnesium oxide, MgO, copper sulfide, CuS. If two elements form more than one compound, then, for example, we have carbon monoxide, CO, and carbon dioxide, CO_2, water which is hydrogen oxide, H_2O, and hydrogen peroxide, H_2O_2. Chlorine and oxygen form four different compounds named according to the number of oxygen atoms in the molecule. When a metal has more than one valence, the compounds it forms with the smaller valence number have the suffix -ous, and those with the higher valence form have the suffix -ic. Cuprous oxide is Cu_2O, and cupric oxide is CuO.

When three elements are combined, no matter how many atoms, the same general rule is followed, except that oxygen is usually left to the very end and hydrogen is usually put at the very beginning. Sulfuric acid, H_2SO_4; sulfurous acid, H_2SO_3; sodium carbonate, Na_2CO_3; barium sulfate, $BaSO_4$; silver nitrate, $AgNO_3$. There are four oxy- acids of chlorine: hypochlorous, chlorous, chloric, and perchloric acids.

A great many salts occur in nature as multiple molecules in highly complicated configurations, often with water molecules hitched on to them. They are always written in formulas with the metallic elements first, the transition metals next, and the non-metals last, but instead of having all the elements in the name, they are called by Latin or colloquial labels that are sometimes as old as antiquity. Cryolite is Na_3AlF_6, and hydroxyapatite is $Ca_{10}(OH_2)(PO_4)_6$. Fool's gold is pyrite, FeS_2.

Several of the groups have identifying names. We already know the inert gases of group O. Group IA elements are called the alkali metals, and group IIA the alkaline-earth metals. There are also the rare-earth metals, which all fit into the box occupied by lanthanum. One of the best known group names is halogen for the group VIIA elements, fluorine, chlorine, bromine, iodine, astatine. Binary compounds of the halogens are called halides, and they are as important as oxides.

We must now examine in greater depth the mechanism of chemical reaction.

15

Reaction in Dynamic Equilibrium

Reaction is energy transfer

WE have already pointed out that some reactions are reversible and some are not, but that all equations must balance. The balanced but irreversible equation for an explosion that wrecked a house is:

$$CH_4 + 2O_2 \rightarrow CO_2 + 2H_2O + E_{heat}.$$

If molecules of methane gas, CH_4, and oxygen are constantly colliding but nothing happens, it must mean that mere shock of collision at ordinary temperatures will not dislodge these particular electron arrangements. But if the colliding molecules are packed with extra energy and the electrons are strained outward against the pull of the nucleus, then a collision will jar them about and they will seek a new configuration which will reduce the tension as much as possible.

Any atom, including those of the inert gases, can be stripped of all its electrons by sufficiently high heat, or intense radiation, or bombardment by superfast particles, or exposure to powerful electromagnetic fields. That is, an electron can be excited to various degrees of energy from its ground, or lowest, state, within the particular configuration around a specific nucleus. Before a chemical reaction can take place the reacting atoms' electron clouds must be ready to rearrange themselves. Ordinary heat of a Bunsen burner, ordinary electrical currents, ordinary radiations will raise the energy of the valence electrons sufficiently to strain the energy balances within the atom to what is called the activation level.

In the kitchen stove when the gas is turned on, a lighted match or a spark will heat the nearest methane (CH_4) and oxygen (O_2) molecules to activation levels. As they collide they react, forming covalently bonded H_2O and CO_2, and in the process enough heat is liberated to raise the nearest molecules to activation level. These combine, liberate more heat, and a chain reaction proceeds.

Since all the reactants and products are gases traveling, at ordinary temperature, at about the speed of sound, all the gas mixture in the room will be raised to activation level in a split second and all the molecules will combine almost at once, with a resulting release of an enormous quantity of heat, which expands everything still more. Even in subzero weather a match flame will start the chain reaction. Of course, if it is applied immediately to the holes from which the gas is escaping, the flame started there will remain there, activating the new gas being fed it continuously.

The point to remember is that the reactants will not react until their valence-shell electrons have been excited to activation levels of energy. Thus, when the reaction takes place, each atom rearranges its electron configuration in such a way that the final result will contain less energy than the original had. Such is an exothermic process, which means literally a process giving off heat.

In an endothermic process, after the energies of the reactants have been raised to activation levels, more energy is needed to push them into reaction because the total energy associated with the product configurations is greater than that for the reactants. An endothermic reaction will never take place, even at activation levels, unless there is available a steady source of energy from the outside to keep pushing the electrons into an arrangement that requires more energies within the atom to maintain. Obviously, if one side of an equation releases heat, the other side must absorb it. Thus if we imagine a reversible reaction, the direction in which it will proceed at the activation level will depend on whether energy is being drained from the total mixture of substances, or is being supplied to it.

In practice many reactions are difficult to reverse, except in a roundabout way, and certain types of reaction have never yet been reversed by man. For instance, the burning of wood produces water and carbon dioxide. Until now no amount of manipulation has reproduced the original wood—or candle, or gasoline, or any natural fuel—from water and carbon dioxide. Persistent attempts recently have produced some molecules which are the basis for wood fiber, but the experiments are not reproducible yet, and what happens is not understood very well. If chemically we disintegrate the cells of once-living organisms, we do not know how to re-create those cells out of the end products, water and carbon dioxide. Only plants are able to combine CO_2 and H_2O into sugar and starch and cellulose, using light energy from the sun, with the help of chlorophyll. Thus the creation of vegetable matter is an endothermic process which absorbs the energy of the sunlight and packs cell molecules with it. Herbivorous animals eat plants and break down the vegetable molecules, liberating their energy, which they use to produce their own complex protein molecules. The carnivorous beasts and man eat flesh and break down those protein molecules to obtain the energy they need to build their own flesh and bone cells.

But a very large number of reactions that chemistry is concerned with, and that industry uses, are reversible. This means that when we write an equation for an actual reversible reaction we are really listing the presence, in a moment of time, of all the reactants and all the products. A mixture of ingredients on both sides of the equation must exist in any case, in any reaction, until one side or the other is completely transformed or removed. We have not pointed out this fact before because our focus was on the process of transformation.

Before we examine this mixture of reacting and product molecules, we must emphasize the obvious fact that no reaction can occur unless the molecules make actual contact. Reactions between gases are therefore the easiest to imagine, and those between liquids that mix are almost as easy. Reaction collisions between a liquid and a gas, a liquid and a solid, a gas and a liquid or a solid, are obviously restricted and less frequent, and are more difficult to arrange. On the other hand, gases are quite difficult to handle, especially in bulk, liquids are less difficult, and solids are not at all difficult to handle. These are real considerations when huge quantities are being processed, and hence affect the design of every factory. A further factor in manufacturing is that a great many substances do not seem to be reactive unless water is present. But this and other considerations we can ignore while examining the general theory of reactions.

Rates of reaction

For the sake of generalization we may symbolize a reversible reaction thus:

$$A + B \rightleftarrows C + D + E_{heat}.$$

For the sake of simplicity imagine that these four molecules are all gases. We begin with a measured amount of A and B in an empty container. The molecules collide billions of times a second, and some few of these collisions may occur between enormously fast A and B molecules, fast enough so that a reaction results from the impact and a few C and D molecules are produced.

If we raise the temperature of the container by heating it from the outside, the number of A + B reactions increases in two ways. First, the average speed of every molecule is raised and the number of collisions per second and the pressure of the impacts is increased, causing a larger number of reactions. Second, since the energy level of all the molecules has been raised, a larger number of them will be at activation level. We must remember that there are always both slow and fast molecules, and molecules with high and low energies.

At a certain temperature the average energy of A and B reaches the activation level and the reaction suddenly proceeds fast. The number of A and B molecules steadily decreases and the number of C and D molecules, of which there were none at the start, steadily increases.

In reality we cannot count the molecules but we can sample the mixture at any moment and with analytical methods weigh the amount of any one of the compounds present in the sample. As we know the whole weight of the sample, the percentage of the different molecules present can be calculated. This percentage is called the concentration.

The apparatus can be arranged so that we can measure the percentage of A, for example, at any temperature, at any moment, and a moment later again, and thus we calculate the amount of A that is being used up, per second. But such a figure represents speed, and it is called the speed of the reaction. The speed, or more properly the rate, of a reaction is the actual weight of A and B being used up per second.

Let us look at the other side of the equation. C and D are being produced, together with a predictable quantity of heat, each time an A-B collision is fruitful. As the reaction proceeds, the number of C and D molecules increases. If the heat produced is kept in the system, then some C and some D molecules will be capable of repacking the available heat energy into themselves when they collide, and they will recombine to produce A and B. The more C and D there is in the container, the more frequent will be collisions between them, the more often they will react.

Just as with A and B, we can measure the actual weight of the C and D molecules that combine per second. The rate of the reaction toward the left in the equation will be nothing at first, but if the heat is not drawn off the rate will gradually increase as C and D increase. At the same time the rate toward the right between A and B decreases. Obviously a time will come when the rate of reaction to the left equals the rate of reaction to the right.

There then will be established a dynamic equilibrium between the two reactions, if the temperature is kept steady. All four kinds of molecules will be present in the container, and the actual number of each will be constant, so that the percentage of each compound in the mixture will not change. Plainly, the actual atoms in these four compounds will be constantly changing places, since the collisions among the four compounds never cease and the reaction in both directions never stops. This is why the equilibrium is called dynamic.

At equilibrium at any temperature, no matter how many of each kind of molecule are present, the number of fruitful collisions between A and B must equal the number of fruitful collisions between C and D, each second. At any temperature, therefore, we can express the concentrations of the molecules in equilibrium in such a way that they equal a constant. This means that if one of the concentrations is changed the concentration of all the others will change proportionately.

A very simple analogy is the decision of a dance committee to allow everyone to come alone, but never to allow more girls than men, or men than girls, on the dance floor. At all times fifty couples, no more, no less, must be dancing. As soon as a girl steps on the floor another girl must get off. If a man gets off, another man must take his place. The partners are constantly changing, but the concentration of men to women remains 50 per cent.

The constant figure calculated from the concentrations in chemical equilibrium varies with the temperature, and it must be calculated for each temperature from observation. It cannot be arrived at theoretically. It is also qualified by the specific manner in which the collisions occur in order to be fruitful, whatever the temperature. Some molecules must hit each other at certain angles in order to react, while others will react on almost any kind of contact. This results in a statistical number: for example, out of 1000 collisions between two specific molecules, only 50 are fruitful.

We may note here this unlooked-for proof that no molecule is a perfect sphere and that most have complicated faces in which the stresses vary, so that some molecules for certain reactions must be hit only on certain atoms within the whole structure. We must not ignore the fact that molecules are always in complicated spins as a result of their nonspherical shape.

Completing a reaction

If we imagine dynamic equilibrium in a reversible reaction, and then imagine adding more A molecules, plainly the rate of $A + B$ will increase for a while until enough more C and D are produced to raise the $C + D$ rate; eventually the two rates will be equal again. If we imagine removing some D, then plainly the $C + D$ rate will drop and so a little more A and B will shift their configurations and get used up, until the concentration of A and B has been lowered a bit and the concentration of C and D has been raised a bit, enough for the two rates to be equal again.

If we imagine removing either C or D completely, A and B will simply continue to react until they are all used up. By removing either C or D as soon as it is produced, we prevent an equilibrium from being established, and the reaction goes to completion, which is the jargon phrase for such an event.

For example, if A and B and C are all liquids being constantly mixed, but D is a gas that is allowed to bubble away, or D is a solid that settles out of the solution, then A and B will continue to react until they are all used up, since no $C + D$ reaction is possible.

In any industry the goal is to obtain as much of some desired product as possible or, to use more jargon, the aim is to get the highest possible yield from the reacting compounds. If the reaction happens to be a reversible one, the removal of any one of the products will accomplish this, but if this is difficult, then forcing into the system an enormous excess of one of the reactants will also drive the reaction to completion. If we have a limited amount of A, say, an ore, but unlimited amounts of B, say, atmospheric oxygen, which costs nothing, or water, which costs very little, then large amounts of B will use up all of A, and we will end with as much C and D as the original amount of B will provide.

Pressure can be used effectively on the whole system when the gases produced are less in volume than the gases used. If C and D occupy more volume than A and B, pressure will drive the reaction to the left.

If the temperature is changed the rate of both reactions will be changed, and changed differently, and a different constant will be established. It is possible that raising the temperature at first increases the $A + B$ rate but that, above a certain point, raising the temperature still more will increase the $C + D$ rate more than the $A + B$ rate. Thus raising the temperature past that point will increase the concentration of A and B. There is an optimum temperature at which A and B have the lowest possible concentration, and it is this temperature that is sought in industry when a reaction cannot be driven to completion. It cannot be calculated theoretically but must be determined by observation.

The role of a catalyst

There is a final consideration concerning the rate of a reaction which introduces a most important field of study: catalytic action. A catalyst is a substance that affects the rate of a reaction enormously without itself being changed. Almost always catalytic action means a speeding up of the rate, and almost always this is what is sought, but there are also inhibiting catalysts that reduce the rate drastically, even to stopping the reaction altogether. Many reactions that are hardly perceptible because they are so slow will accelerate to almost explosive speed with the right catalyst, and thus the study of catalysis is vital to many industries that could not be conducted profitably if some one reaction in a whole series ran at its natural pace.

One of the most fascinating fields of research concerns the catalysts that every living organism must manufacture in order to exist. We have mentioned that green plants combine carbon dioxide and water into chains of starch and cellulose in an endothermic process requiring vast quantities of energy in the form of sunlight. A little reflection reminds us that CO_2 and H_2O are always present in the air, and so is sunlight—yet nothing happens in the air. In contact with a catalyst called chlorophyll, the reaction between these molecules proceeds at the brisk rate of growing leaves and grass. Without chlorophyll no leafy plant could exist, and without green plants no animals could exist, because they cannot utilize the energy of the sun for their own metabolism.

The mechanism of catalysis is still somewhat speculative, although the theory has been narrowed to two possibilities. A molecule striking a catalyst—also a molecule or atom or ion, though not necessarily at molecular level of dispersion—may be seized by it and held in such a way that its structure is distorted into the configuration it would have at activation level; this strained molecule will react when hit by the other reactants. The other possibility is that the catalyst actually forms an intermediate compound with one of the reactants, which is then vulnerable to the other when they collide. The point is that at the end of the reaction the catalyst reappears exactly as it was at the start, with not one atom of it used up or changed in any way.

A fixed amount of catalyst in the reaction chamber, therefore, will bring about the speedy reaction of endless amounts of reactants. The still puzzling aspect of the phenomenon is that only very small quantities of catalyst need be present. A pinch is enough for many pounds. Since some catalysts are fairly expensive—finely divided platinum being a very popular one—the search for catalysts never ends.

(NOTE: a catalyst affects the rate of a reaction. It does not and cannot ever bring about a reaction that is not already happening. It merely speeds it up enormously as though the reactants had been somehow excited way beyond the energy levels that temperature alone could produce.)

A catalyst does not speed up only one side of a reversible reaction. Both sides are speeded up equally. The equilibrium constant is not changed by the addition of a catalyst. The concentrations of the substances when equilibrium is reached, at any temperature, is the same as it would have been without the catalyst.

The law of mass action

To sum up, then, when a reversible reaction is at equilibrium, the concentration of the reactants remains constant. Any change in the environment or in the amounts of the ingredients will upset the equilibrium, but the rate will then shift one way or the other in such a manner that the equilibrium will be restored. This is called the principle of Le Chatelier, and it is one of the most important fundamental laws that science has uncovered. Its implications are deep, including the assertion that no static conditions can exist in the universe.

The law of mass action derives from this principle and it puts the concentrations of the reacting materials into a simple formula. For instance, for the reaction

$$A + B \rightleftarrows C + D$$

the formula is

$$\frac{[C] \times [D]}{[A] \times [B]} = K_T$$

where the square bracket indicates concentration, and K is a constant at the temperature T. If we change the concentration of any one ingredient, the concentrations of all the others will change automatically in order to maintain the constant K. All the discussion in this chapter is contained in this formula. A catalyst does not change it but only brings equilibrium faster than the temperature T can.

Any new industrial process is worked out first in the laboratory to find the equilibrium constants at different temperatures, according to the law of mass action. A pilot plant is built to handle bulk quantities of materials and to study the effect of large surfaces and of pumps and other kinds of machinery on the reaction. An unlooked-for and small fluctuation of temperature or pressure at some point in the assembly may reveal a totally unexpected change in the rate of one of the reactions. It might pay to install an expensive purifier, or drier, or pump, or to redesign a bubble tower, or a scrubbing operation, just to reduce by a fraction the presence of some intermediate product. The final installation of production equipment will usually accumulate many bugs related one way or another to the equilibrium of some reaction, whose yield changes as a result and affects the final product.

The living organism inevitably is seen more and more in the light of this reality. A slight alteration in some part of the body's chemistry will produce grave changes elsewhere through an intricately linked chain of dynamic balances.

Le Chatelier's principle goes against many everyday experiences that can be defined as illustrating Newton's famous law that to every action there is an equal reaction. If I hit the table ten times as hard, my fist will smart ten times as much. We tend to think of our encounters with brute matter and even with one another in terms of thrust and counterthrust, blow and return blow, an eye for an eye. Much of life is indeed a stiffening against challenge and threat, but the point is that we are not in dynamic equilibrium with the table, or with a brick wall we pile our car into, or with most human beings we encounter. Le Chatelier's principle refers exclusively to a system whose parts are in dynamic equilibrium. Then, a pressure, a blow, an intrusion will create a response in that system which will absorb the intrusion, eliminate the pressure, and restore equilibrium.

Obviously any system can be overloaded beyond its capacity to regain equilibrium, and this point of failure is also of interest to all the life sciences. In recent years a new science called ecology has evolved around the concept of a system of organisms in equilibrium with one another and with their environment.

The parts of a living cell are in equilibrium not only with one another but with what happens outside the cell too. The cell constantly adjusts to its environment and will not react as a brick wall to missiles hurled against it. The exchange of energies never stops but flows in such a way as to preserve the system. But it is very simple to overload and destroy the cell.

The concept of feedback grows partly out of what we have been discussing. Machines that correct their mistakes, or regulate themselves against changing loads, and especially the parts of electronic computers, are all systems in dynamic balance, capable of readjusting to a new equilibrium when the old one is disturbed.

16

How to Make Predictions about
Chemical Bonds

Atom and ion sizes

IT should no longer surprise the reader to be told that the general
concept of chemical change is changing. Alchemistry was stabbed
to death by Lavoisier's gravimetric procedure, and the corpse was
buried by Dalton's atomic theory. But the vocabulary of chemistry
has retained "transformation" and "change" and other meaningful words for the abstract idea of metamorphosis, reaction, and so
on, and has retained an insistence that chemical changes are fundamentally different from physical changes. Even after chemical
reaction was proved to be electromagnetic in nature, the physical
properties of the atom were considered to be something quite apart
from the chemical properties.

This distinction between physical and chemical behavior, however, is steadily diminishing, and we now focus on an aspect of the
atom that we have only casually referred to: its size relative to
other atoms. The actual size of an atom has been the target of
many research efforts, almost none of them connected with chemistry, and one cannot see that it will ever seriously enter into chemical work, any more than the actual weight of atoms does. But
knowledge of the relative size of atoms is becoming increasingly
useful.

The geometric precision with which atoms line up in crystals
has fascinated scientists for thousands of years, but only the atomic

and molecular theories could penetrate those smooth, glistening surfaces and perfect edges. Finally X rays and streams of electrons, poured through a crystal, cast patterns of light and shadows on photographic films, from which the latticework of atoms in a crystal could be worked out.

It became clear that the relative size of atoms had a great deal to do with their specific arrangement in a crystal. The geometry of the lattice was not wholly the result of electron and nuclear tensions. Electron clouds had to be touching one another's peripheries as the atoms lined up, and crystal shapes composed of different-sized atoms, therefore, reflected the fact that small atoms could nestle in between large atoms. Atoms of the same size could not pack together as densely.

Most crystals of interest are not pure elements but compounds, and in many crystals the particles are ions linked by their charges. Thus the size of ions can be determined too. The marked difference found between the size of an atom and that of its ion stimulated further work in this field, and today we can compare the two with reasonable confidence in our accuracy. The diameters of atoms and the diameters of ions both reveal a remarkable periodicity, with which we can explain some of the chemical properties of the elements. Having explained chemical reaction as a movement of electrons in order to achieve an inert configuration of 8 electrons in the outermost shells of the atoms, we now enrich this purely quantitative concept with a much more qualitative comparison of actual sizes.

Perhaps it is useful to pause and imagine atoms or ions in a crystal, in perfect orderly arrangement, each cloud of electrons pressing against the surrounding clouds, and at the heart of each cloud, a far distance in from the valence shells, the tiny, positively charged nucleus with practically the whole of the atom's weight in it. An electron is very small compared to the whole atom, and if it is traveling freely and fast it may pass unhindered through an electron cloud and through the space between the cloud and the nucleus, and if it hits the nucleus it will bounce off and change its direction, cutting back through the cloud and slicing straight ahead through other clouds. The electron is not seized by the positive nucleus because the nucleus already has its complement of electrons. A slowly moving electron would not be able to pass through the repelling force of the cloud around a nucleus, but it could move about between the clouds. Each electron in a crystal belongs to some nucleus, but it can be detached, and at all times it is also subjected to the attraction of all the nuclei in the vicinity, and to the repulsion of all the other electrons in the vicinity.

A negative and a positive charge will attract each other four times as much if the distance between them is halved. An electron will repel another four times as much if the distance between them is halved. Therefore the distance between a nucleus and its electrons is determined by the force of attraction with which each electron is held, and the distance between the electrons themselves is determined by the force with which each electron strains against its neighbors. Many of these forces have been computed from a study of crystal structure and from a broader aspect of this subject called the solid state, which considers other conglomerations of atoms and ions besides those in crystals.

We know from experiment that the diameter of an atom has an influence on the energies of the valence electrons, and we know exactly how strong that influence is. This is not theory.

As the number of protons increases from element to element, each extra positive charge adds enough strength to the attractive force of the whole nucleus to pull a little closer to itself all the electrons. They are pulled in against the forces of repulsion among the electrons which keeps them apart. Thus, the diameter of the atoms of the elements in Period 2 shrinks steadily from lithium (Li) to fluorine (F), which, with six more electrons in its outer shell than lithium has in the same shell, is actually less than half as large. But then neon (Ne), with one more proton and one more electron, suddenly swells up. Apparently, to get that eighth electron into the L shell, all eight have to be pushed away from the nucleus.

The next element, sodium (Na), with a single electron in a new shell, is larger than neon but not quite as much larger as one might expect. The larger sodium nucleus has pulled its electrons a little closer to itself. Magnesium (Mg) is smaller than sodium, aluminum (Al) is smaller than magnesium, and the diameters decrease as they did in the period above, element by element, until argon (Ar) is reached, which once again expands to accommodate its eight electrons in a perfectly balanced outer shell.

The pattern of shrinking is repeated within each period.

As for the ions that are formed by the atoms, the metals on the left side of the table lose electrons, and hence we would expect their ions to have smaller diameters than their atoms. And so they do, but the difference is more than we might expect. What happens, as we pull off the valence electrons, is that the nucleus with its now unbalanced positive charge is stronger still than it was in the neutral atom and will pull the remaining electrons closer to itself.

The nonmetal ions, however, gain electrons, and thus contain more electrons than their atoms do. We would expect these ions to be larger than their neutral atoms, and they are larger, but the difference again is more than we might expect. Apparently to squeeze in enough valence electrons to have eight, the outer shell must be enlarged considerably. In fact the diameter of the chlorine ion is larger than the diameter of the following inert atom, argon (Ar), just as the ion of fluorine (F) is larger than the atom of neon (Ne).

The size of atoms going down in any one group naturally increases. Lithium (Li) is the smallest atom in group IA, and cesium (Cs) is the largest. Information on franconium (Fr) is not complete, but presumably it will prove to be larger still. Fluorine (F) is the smallest atom and iodine (I) the largest in group VIIA. So the size of atoms decreases going from left to right in a period but increases going down a group.

The largest atom is therefore at the bottom of group IA, and the smallest, apart from hydrogen, at the top of group VIIA. Hydrogen is the smallest of all. The smallest ion is at the top of group IA and the largest ion at the bottom of group VIIA.

The pattern reveals still more reason for the periodicity of properties among the elements. The willingness of the nucleus to release, to accept, or to share its valence electrons can be predicted by studying the relative sizes of the atoms and of their ions, and this willingness has been given a name: the electronegativity of the element.

The strength with which the nucleus holds its electrons varies according to the number of protons in the nucleus, the distance between the nucleus and the electron, the number of other electrons in the same shell, the number of electrons between the one we are interested in and the nucleus, these latter shielding valence electrons from the pull of the nucleus. Electronegativity, as the term is defined, is the force required to pluck an electron from its place in the atom, whether that electron is in a neutral atom or in an ion. Electronegativity is therefore highest with those elements that do not form ions at all, the inert gases. The next highest electronegativity would be found with those elements most eager to complete their octets—the group VIIA elements. Electronegativity would be lowest with those elements most ready to give up valence electrons—the group IA elements. It takes less effort to remove an electron from a metal atom than to remove one from a nonmetal atom. Another comparative term for the same property is "electron affinity," but "electronegativity" seems to be winning out.

No such thing exists as a basic unit of electronegativity. It is a comparative value, and so we must establish a standard. The least electronegative element is at the bottom of group IA, the largest of all the atoms, with its valence electron farthest from its nucleus. Electronegativity increases from cesium to lithium. It increases in every group from bottom to top because the smaller atoms in each group cling to their electrons with greater force than the larger atoms. Electronegativity increases going to the right in any one period because the atoms get smaller as the number of protons increases. Thus fluorine is the most electronegative of all the reactive elements. Fluorine was given an electronegativity of 4, and the electronegativity of all other elements was compared to this.

Diagraming chemical bonds

We can now add some fresh information to what we already know about chemical bonds.

In Chapter 13 we named three quite different modes of bonding, two of which were defined. In one, electrons actually transfer from a metal to a nonmetal; in the other, electrons are shared between nonmetals. If atom B is far to the right in the periodic table and atom A is far to the left, then B is much more electronegative than is A. B will actually seize electrons from A to put them into its own valence shell, and A willingly lets them go. The resulting cation and anion will cling together in electrovalent bond. If we represent each electron in the valence shell by a dot, then we can pictorialize what happens between two elements whose electronegativities are much different:

$$A \overset{\cdot}{\cdot} \ + \ \overset{\cdot\cdot}{\underset{\cdot\cdot}{B}}\!\!: \ \rightarrow \ A^{++} \ + \ \overset{\cdot\cdot}{\underset{\cdot\cdot}{:B}}\!\!:^{--}.$$

But if the difference between electronegativities is slight, if the two elements have more or less the same grip on their electrons, then only a covalent bond can be formed. They both need electrons with the same urgency, that is, they both have room for more electrons at the same energy levels, and so they will share with each other the ones they already have, in order to complete their octets. With a coordinate covalent bond the picture is much the same. Using dots for valence shell electrons, we can pictorially set up the equation:

$$\overset{\cdot\cdot}{:A\cdot} \ + \ \overset{\circ\circ}{\underset{\circ\circ}{{}^{\circ}B}}\!\!\circ \ \rightarrow \ \overset{\cdot\cdot}{:A}\overset{\circ\circ}{\underset{\circ\circ}{{}^{\circ}B}}\!\!\circ.$$

Now let us select two nonmetals that have different electronegativities. The first one energetically seizes electrons to become an ion, and the other rather reluctantly, but both take up, rather than give, electrons. If these two react, then they will form a covalent bond in which they will share the electron pair unequally. The first one will pull the shared pair closer to itself because it has a higher electronegativity. The other, being weaker in this respect, will allow the shared electrons to move farther from its nucleus. The shared pair will nestle into the first one's cloud.

Using the dot system:

$$\overset{\text{oo}}{\underset{\text{oo}}{A}}\text{o} \;+\; \overset{\text{..}}{\underset{\text{..}}{:B}} \;\rightarrow\; \overset{\text{oo}}{\underset{\text{oo}}{A}} \;\; \overset{\text{..}}{\underset{\text{..}}{\text{o:}B}}.$$

In this covalent molecule AB the valence electrons are more closely associated with B than with A. We can say that B has pulled more of the negative charges into its space in the molecule, leaving the rest of the space somewhat deficient in electrons, and therefore a little more positive. The molecule as a whole is still neutral because the total number of electrons equals the total number of protons. But the more electronegative end of the molecule will be more negative than the other end.

Such a molecule is called polar. The distribution of electrons in a polar molecule makes one end of it more negative than the other. It has, in other words, poles of electricity and behaves quite like a little bar magnet. It will align itself in an electrical or magnetic field with its negative end pointed at the positive pole. It will also accommodate itself to another polar molecule, swinging, when it can, so that they touch their oppositely charged ends together.

In a polar molecule the bond is covalent, but the pull on the shared pair is unequal. If the difference between the electronegativities of the two atoms is quite large, then the shared electrons will be held so closely to the stronger that the situation is very nearly ionic. The stronger atom has very nearly removed the other's electron.

It is therefore possible to say that all chemical bonds are basically the same and that, depending on the actual difference in the electronegativities of the combining atoms, the bond will be electrovalent, or covalent, or polar.

Here are some of the electronegativities of the elements: lithium has a value of 1, while chlorine has 3. The difference between them is 2, and this is sufficient to detach lithium's electron completely and transfer it to chlorine. Boron has 2, and so the difference between it and chlorine is only 1. The two elements can bond only covalently, but chlorine will hold the shared pair closer to itself, to form a polar molecule. Chlorine and nitrogen, with a value of 3, will form covalent, nonpolar compounds.

Obviously there must be some uncertain situations. An element of middling electronegativity will seize electrons from those having very low electronegativity but will yield its own electrons to those having a very high electronegativity. Such elements would sometimes behave as metals and sometimes as nonmetals, depending on the element with which they combine. These are the transition metals, which appear in the center groupings of the periodic table. To give them a more distinguished name than transition metals would violate the simplicity of metal and nonmetal categories. There are twenty-two elements which behave as nonmetals at all times, twelve elements are always clearly metallic, and all the others are transition metals, or borderline cases. This categorization is achieved largely on the basis of electronegativity values.

The whole system of valence counting can now be revised. There used to be a covalent number and a negative electrovalent number for all nonmetals, a positive electrovalent number for all metals, and a variety for transition metals. The concept of electronegativity clarifies many chemical reactions that were quite inexplicable and diagrams many formulas whose bonds could not be otherwise understood. Today we classify all chemical reactions according to the assumed movement of valence electrons from less to more electronegative atoms, or ions, or molecules, or particles, or objects with electrical charges on them. Chemistry and electricity are inseparable.

17

Variations on the Theme of Fire

Oxidation

THERE is a type of reaction that is probably the oldest chemical manipulation in man's history, but whose modern name is only as old as Lavoisier: oxidation. The best known examples are the burning of natural fuels. Wood, coal, oil, fats all combine with the oxygen of the atmosphere at such a high rate that the heat evolved creates a flame. The structure of a flame is far more complicated than it would seem to be but, broadly speaking, the oxidation of the surface layers of fuel liberates enough heat to vaporize the next layers of fuel before they oxidize. These inflammable vapors rise to mix with the air and then burn in the shape of a flame, within parts of which there is almost no oxygen, while within other parts there is an excess of oxygen. The hot products and unused hot air rise above the flame.

The same oils, fats, wood, and coal will combine with oxygen merely on exposure to the air, but very, very slowly—so slowly that the heat of the reaction is dissipated at once. If the heat of the reaction is not allowed to escape, as in a bunch of oily rags stuffed in a barrel where oxygen can still get at it, enough heat may build up from a few reactions with oil molecules at activation level to raise more molecules to activation level, and thus starts a little chain reaction. The heat collects in the enclosed space, and eventually the oily rags burst into flame. The phenomenon is called spontaneous combustion.

When finely milled wheat flour floats in the enclosed, still air of a warehouse, each tiny kernel of flour oxidizes slowly with a tiny evolution of heat, but millions of those kernels sometimes release enough total heat to raise the temperature of the mixture to kindling point and then it explodes. This is another example of spontaneous combustion.

Metals combine with oxygen at ordinary temperatures in familiar ways. Iron rusts, and aluminum, lead, and other metals acquire a dull film, and some metals, such as magnesium, combine so fast with oxygen that, once the reaction has begun, they burst into flame. Some metals—sodium, for example—have to be kept in kerosene or some other sealer against contact with oxygen.

Almost every element combines with oxygen under specific conditions, and hence oxygen appears in more inorganic compounds than any other element.

Thus oxygen is associated with a very large number of reactions. In the old notation, oxygen's electrovalence was -2 and its co-valence was 2. With metals the general formula of an oxide is $M^{++}O^{--}$, for instance, calcium oxide is $Ca^{++}O^{--}$, or sodium oxide, $Na_2^+O^{--}$. With chlorine, oxygen forms several covalent compounds, one of which is ClO_2. With many compounds oxygen forms polar molecules, the oxygen end being negative because, except for fluorine, oxygen has the highest electronegativity. Water is perhaps the most important compound we know, and it is polar.

In every case when oxygen combines, except with fluorine, oxygen gets electrons in a covalent or electrovalent or polar bond. In trying to get away from the mixture of valence numbers, chemists decided to enlarge the concept of oxidation and to include under that term any reaction at all in which electrons are gained by an "oxidizing" agent, which could be an atom, an ion, or a molecule, whether or not oxygen was involved.

Thus, in the combining of sodium and chlorine, chlorine is an oxidizing agent because it gains electrons; sodium, which loses electrons, is oxidized by chlorine. The positive electrode of a battery is also an oxidizing agent.

What shall we call the opposite of oxidation? The opposite of an oxidizing agent would donate electrons.

Reduction

There is a group of reactions which have been known since the Egyptians discovered how to smelt copper from its ores. When the ore is crushed and mixed with charcoal, and heated, pure molten copper runs out, leaving slag behind. This process, which can be used with most common metals, was called reduction by the alchemists, carbon being called the reducing agent, and it still has the same name in metallurgy. A general formula can be written for the process of reducing an ore, which is either an oxide or a compound that can be converted to an oxide:

$$2 \text{ Metal}^{++} \text{O}^{--} + \text{C} \rightarrow 2 \text{ Metal}^{0} + \text{CO}_2\uparrow.$$

That is, the metal ion in electrovalent bond with the oxygen ion is forced to accept electrons and become the neutral metal atom, while the oxygen ion loses its two extra electrons and bonds covalently with carbon to escape as carbon dioxide.

In this reaction a metal gains electrons instead of losing them, and is liberated from its oxide, a process opposite to oxidation, hence, when a term to balance oxidation was sought, the word "reduction" was chosen. The reducing agent is any ion, atom, molecule, or object that yields electrons. When sodium and chlorine combine, sodium is the reducing agent. The negative electrode of a battery is a reducing agent because it supplies whatever comes along with electrons.

Thus, every oxidation reaction is also a reducing reaction: the oxidizing agent gets reduced as it oxidizes the reducing agent.

To gain electrons, even in the depths of some complicated molecule, is to be reduced. To lose electrons is to be oxidized.

Oxidation numbers

We can now devise a new way of reporting electron movement in chemical reactions—by referring to the oxidation state of any atom. A neutral atom has an oxidation state of 0. A negatively charged ion is an atom in a minus oxidation state, and its oxidation number would be the number of negative charges. Oxygen in a compound, whether electrovalent or covalent, is always in an oxidation state of —2; thus when it combines with anything it lowers its oxidation state from o to —2. A metal when it combines raises its oxidation state from o to +1 or +2 or +3.

The point to keep in mind is that these words are used in a strict manner and have no significance other than the count of electron shifts. All textbooks use the phrase oxidation-reduction, or redox, and discuss oxidation states and oxidation numbers, and even popular articles employ them as a matter of course. Valence numbers are used less and less. Once again a concept has shifted from a static to a dynamic interpretation of facts.

Valence electrons are no longer seen as poker chips won or lost according to set rules. They are items of energy count which flow wherever the slope of forces indicates. Once valence meant hooks on the atoms, and it is fruitful to compare the introduction of valence in Chapter 8 with the picture we have now of every atom or ion seeking a lower or higher oxidation state, determined by the relative inner tensions of all the atoms involved. We can replace all the valence numbers by oxidation numbers, as follows.

The oxidation state of oxygen in all but a few compounds is always —2; of hydrogen it is always +1. Both these numbers are the same as the old valence numbers. The valence of sulfur is —2 but it forms two compounds with oxygen, sulfur dioxide, SO_2, and sulfur trioxide, SO_3. The total oxidation number of oxygen in the first is —4; therefore the oxidation number of sulfur must be +4. In the trioxide, the total oxidation number of oxygen is —6, therefore the oxidation number for sulfur must be +6. When the dioxide is turned into the trioxide, the oxidation state of sulfur is raised from +4 to +6; the sulfur has been oxidized by the addition of one more oxygen atom.

The formula for sulfuric acid as determined by analysis is H_2SO_4. Hydrogen has a total oxidation number of +2, and oxygen has —8. The sum of these is —6. Therefore sulfur in this molecule must have an oxidation number of +6.

The sum of all the oxidation numbers in a formula must always equal 0, or, if an ion is involved, the amount of charge on the ion, and if we imagine that we are counting the actual disposition of valence electrons in different compounds, most elements will turn out to have more than one possible oxidation number. The advantage in using them instead of valence numbers is that complicated formulas can be untangled. Not everything is known about electron bonding, whether ionic or the sharing of pairs, and quite obviously this ignorance is reflected in the several attempts chemists have made to arrive at a simple or altogether reliable system of reporting linkages between atoms. The use of oxidation numbers, which developed from the facts of relative electronegativities, enables us to compute the changes that can't be pursued with valence numbers, even if no diagraming of the changes is possible. Some elements have as many as six oxidation numbers and thus can act as oxidizing or reducing agents in a great variety of reactions, raising or lowering their oxidation state to suit the complex meshing of the different electron affinities involved. Actual oxidation numbers vary from -4 to $+8$, and we simply assume that this many electrons of the valence shells can be involved in the bonding, depending on the element.

All the sciences have shorthand methods to handle frequent and essential problems, and often the methods are merely arithmetical. But they work better than methods that try to account for the whole picture in its variety of details. Simplification is a very genuine goal in all the sciences. In fact, the argument is sometimes put forward that science is nothing but a way of simplifying information. In any case, the real usefulness of oxidation numbers comes to the fore in the balancing of equations which fall into the oxidation-reduction category, one of the two major categories of reactions.

We now turn our attention to the other major category, the acid-base type of reaction.

18

Variations on the Theme of Water

The shape of the water molecule

W E have discussed briefly how the relative size of atoms, which is a periodic function of the atomic numbers, faithfully reflected in the periodic table, determines their arrangement in molecules, partly because of the effect of the distance between the nucleus and its own electrons and those of nearby atoms. And if we also keep in mind the difference between the sizes of atoms and of their ions, and the effect of what we call electronegativity, then we can speculate on the general properties of some simple molecules. The point to be emphasized here is that the periodic nature of the electron structure of the elements, which determines the way in which they form compounds, is not reflected in any comparable periodicity of the compounds. Compounds are merely similar or not similar. They belong to families, to groups, to categories, to types, but nothing quantitative relates the groups to one another in a periodic, predictable way.

Symmetry and asymmetry of shape affects molecular properties. The valence electrons mesh and blend and rearrange themselves around the different nuclei while inner shells of electrons resist disarrangement and the nucleus remains impervious to the changes. The way molecules condense into the liquid state and freeze into the solid state depends on their actual sizes, on their shapes, on their polarities, on the presence of a few key atoms such as hydrogen. A dumbbell-shaped molecule that has positive and negative ends and consists of only a few atoms will have properties related to that construction, and they will be quite different from the properties imposed by a pyramidal-shaped, symmetrical, perfectly balanced molecule, and different from those of a molecule with a long, thin chain of atoms, or of one with rings of atoms attached to it.

We focus on the water molecule, the most important by far of all molecules for life on earth and the most prevalent on the face of the earth. It consists of two hydrogen atoms covalently linked to one oxygen atom, which, being more electronegative, holds the shared pair closer in a polar bond. Water combines with many substances to form complicated molecules which then break down into a variety of classifiable compounds. The process is called hydrolysis and is one of the most important categories of reactions. Many reactions which formerly were considered to be quite simple, modern research has proved to be hydrolysis.

The shape of the water molecule has been determined by countless researchers. In all its states, whether vapor, liquid, or solid, the water molecule is not symmetrical; the oxygen atom is not at the center of a straight line between hydrogen atoms on either side of it, thus: H—O—H. The two hydrogen atoms are positioned more on one side of the oxygen than on the other as in the configuration with an angle of 105 degrees between the two hydrogen atoms.

Each hydrogen nucleus, in reality a single proton, is covered only by the swing of its shared pair of electrons, while the oxygen nucleus has a total count of ten electrons around it. We can conclude that the molecule is not only asymmetrical but has most of its electrons concentrated at the oxygen end. It is a polar molecule. Like a tiny magnet, it has a positive and a negative end.

If we compare the water molecule to others which have more or less the same size and more or less the same number of protons and electrons, but which are not polar, such as methane (CH_4), and ammonia (NH_3), we find that the physical properties as well as the chemical properties of these compounds are radically different from those of water.

Water freezes at a lower temperature and boils at a higher one than the others. Its heat of fusion and heat of vaporization are higher than for any similar compound which is not polar.

If we imagine the water molecules sliding and tumbling in the liquid state at the average kinetic energy represented by the temperature, it is plain that whenever a positive end of a water molecule touches the negative end of another, they will tend to cling, whereas two positive ends coming close will tend to swing away from each other. If a group of molecules are found to have swung into a mutually harmonious pattern of oppositely charged ends touching, the pattern will persist and will act as a brake on the movement of the molecules through the liquid. A rise in temperature will make them swim about a little more forcefully but not as much as if they were not polar. The clinging of positive to negative ends will persist right to the boiling temperature. Not only will a liquid that is polar boil at a higher temperature than a similar liquid that is not polar, but the attraction between polar molecules has to be broken with extra amounts of heat. The latent heat of vaporization will be higher with polar molecules than with nonpolar. If a polar molecule is cooled, the patterned arrangement of the molecules will increase and they will persist in this liquid pattern to temperatures below that which we would predict for the same sort of nonpolar molecule. Water's latent heat of fusion will be higher, too. The viscosity of water also is predictably higher than of comparable nonpolar compounds.

Water, like all substances, shrinks in volume as its temperature is lowered, but just above freezing, about 4° C., it has the almost unique property of suddenly expanding so that the density of freezing water is lower than that of just cold water. When it turns to ice, the ice has a still lower density and hence floats on the surface instead of sinking to the bottom, as do almost all other solids when they form in their own liquid. The weight of a bucket of ice is less than the weight of a bucket of water.

If the density of ice were higher than of water, and ice sank, rivers would freeze on the bottom first. Presumably all the fish would be forced to swim in steadily shrinking amounts of surface water, as the ice thickened from the bottom upward, until they expired flopping on top of the ice. Textbooks often praise this phenomenon of nature: how fortunate for us, they say, that ice floats, or all our fish would freeze into the rivers and lakes. The implication of a careful Mother Nature somehow arranging matters to preserve the charming status quo for fishermen is so contrary to the correct view that it illustrates how difficult it is for the mind to see itself as a product of universal forces instead of as an independent traveler in the universe. We do prefer to see ourselves as tourists here on earth for a while, with our own private destinies, immune from the mundane, pleasing, or unpleasant laws of nature.

The physical properties of the small wedge-shaped polar water molecule explain some of its phenomenal ability to dissolve almost anything. It worms its way among the atoms and ions of a solid and pries them apart, using either its positive or negative end as required. In this statement, too, a common fault of thinking is revealed: the facts are described as though the water molecule went at the business of dissolving other compounds with a private will of its own. The reality is that when the weak magnetic pull of the water molecule acts upon the atoms bonded in the solid, their own bonds will be loosened a little. Many other polar liquids are good solvents for the same reason.

But water has still another ability—to form what are called hydrogen bonds.

The hydrogen nucleus, a single proton, is so small compared to all atoms that it has a few special characteristics, whether we find it alone as a positive ion—in other words, a proton pure and simple—or whether we find it as part of a molecule and refer to it as hydrogen because it is associated with two electrons, constituting the filled-up K shell. One of its properties is that it penetrates the electron cloud of several strongly electronegative elements, oxygen being one. As a result of this penetration one can speak of a weak bond having been formed. The bond is not chemical in any way because no change in electron structures results from it, and therefore it is called, for want of a better word, a hydrogen bond.

Even when the proton is covalently linked into a molecule, large or small, if it dangles in just the right way somewhere on that molecule, it will form weak hydrogen bonds with molecules that contain oxygen, or fluorine, or nitrogen, when their electron clouds are available. Since the discovery of the hydrogen bond in water, its prevalence and persistence in these other compounds has been recognized in an increasing number of reactions, and it is now considered to be an important factor in those reactions, many of which relate to the living organism. Let us restrict ourselves to its role in water.

The hydrogen end of any water molecule digs into another water molecule's cloud of electrons at the oxygen end. Thus hydrogen bonds increase the force with which polar water molecules stick together. And hydrogen bonds make water an even better solvent when it can form weak hydrogen bonds with molecules in the solid state, thus loosening the bonds among those molecules.

It should be noted that a molecule need not be polar in order to form hydrogen bonds, and that a hydrogen bond is not in the category of chemical bonds.

Ionization of water

The most important property of water, however, by far the most important, is its ability to break up of its own accord into positive and negative ions. The explanation of these ions leads us into the modern theory of acids and bases.

Acids and bases are groups of compounds, some of which were known and named in ancient days. Next to fire, or oxidation, they are the most powerful reagents man has for producing chemical reactions. All acids taste sour, attack metals—not all acids attack all metals—and react with all bases to produce salts. Bases taste bitter, have a soapy feel, and combine with all acids to produce salts. Both bases and acids conduct electricity. Acids change certain dyes one color, bases another. Several theories have been used through the centuries to explain acids and bases, but we will restrict ourselves to the most popular one today.

So far, we have spoken of ions as atoms that have gained or lost electrons and thus become electrically charged. But the term "ion" refers to any particle with an electric charge. Thus any molecule or fragment of a molecule is an ion when it has more or less electrons than is required for neutrality by the total number of protons in that molecule or fragment. An ionized particle is simply one that has lost its electrical neutrality. Radical is another name for an ion that is not a single charged atom but a group of covalently bonded atoms with an electrical charge, negative or positive. "Ionized air" refers to nitrogen, oxygen, water, carbon dioxide, and other molecules in the atmosphere that have had a few electrons knocked out of them by the speedy passage of free electrons, say, in a bolt of lightning, or of neutrons hurtling in from outer space, or of high-frequency rays. In the laboratory most inorganic reactions we deal with involve ions, or radicals.

In absolutely pure water the following reversible reaction takes place:

$$H_2O + H_2O \rightleftarrows OH^- + H_3O^+.$$

Using the dot system for indicating valence-shell electrons, we have:

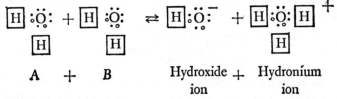

A + B   Hydroxide + Hydronium
                    ion   ion

One hydrogen nucleus, or proton, in molecule A actually detaches itself and migrates to form a coordinate-covalent link with two electrons of the oxygen in molecule B. (NOTE: This is not a hydrogen bond.) The migrating proton has left behind its own electron on the oxygen atom in A, thus turning that fragment into a negative ion or anion. The migrating proton has added its own positive charge to molecule B, turning it as a whole into a positive ion, or cation. In other words, the two molecules of water have reacted to become ions. The radical H_3O^+ is called the *hydronium* ion, and the radical OH^- is called the *hydroxide* ion.

In pure water this reaction is barely perceptible, and water was never considered to be ionized until refinements in electrical equipment detected the flow of a faint current. A current cannot flow unless ions are present. When ions are added to pure water the conductivity increases enormously. Sea water is quite a good conductor, since it is full of ionized salts.

But in pure water the rate of the reaction to the right is very, very low, compared to the rate of the reaction to the left. One can say that practically every collision between hydronium and hydroxide ions results in the formation of water molecules, while only one out of billions of collisions between water molecules results in ionization. The dynamic equilibrium that exists in this reversible reaction keeps the concentration of the ions constant at any fixed temperature, according to the law of mass action.

In a fixed quantity of pure water, as we can see by the equation, the actual number of H_3O^+ must equal the actual number of OH^-. Therefore the concentration of the two ions, or radicals, is the same, no matter how much pure water we examine. In pure water at room temperature the concentration of each ion is measured to be the minute fraction 1×10^{-7} (0.0000001). We can now set up the formula provided by the law of mass action:

$$\frac{[OH^-] \times [H_3O^+]}{[H_2O] \times [H_2O]} = K_T.$$

The constant K varies with the temperature, but the H_2O is always present in such vast quantities relative to the ions that we can consider their concentration a constant at all times. Therefore the product of the concentrations of the two ions is a constant:

$$[OH^-] \times [H_3O^+] = K_T$$
$$[1 \times 10^{-7}] \times [1 \times 10^{-7}] = K_T$$
$$10^{-14} = K_T.$$

The point to keep in mind is that the ionization of water is a reversible reaction which is always in dynamic equilibrium, no matter what the concentrations happen to be. The concentration of one ion multiplied by the concentration of the other ion is always the same constant at a particular temperature, no matter how much water is present and no matter what else is dissolved in that water.

An extremely rough analogy is provided by the equation $a \times b = k$, where k is always 100. Then, for example, if we increase either a or b, we must decrease the other in order to keep the product the same. Thus $20 \times 5 = 100$, or $50 \times 2 = 100$, or $40 \times 2\frac{1}{2} = 100$. As the concentration of either is changed, the concentration of the other will change automatically.

Acids

Suppose that the hydronium ion concentration in pure water is somehow increased, perhaps by direct addition of the ion. The number of fruitful collisions between the two ions will increase, and the rate of the reaction to the left will increase. Hydroxide ions will be removed until the product of the two is again the same constant. In this way, by simply adding hydronium ions to water, we can reduce the amount of hydroxide ion to a negligible amount.

A solution in which either ion is in excess of the other is no longer pure water. If the H_3O^+ is in excess, the solution has a sour taste, turns a dye called litmus a red color, and will react with many metals to produce hydrogen gas. It will also react with alkaline substances to produce salts. But these are precisely the characteristics of a large group of compounds called acids. In other words, a water solution in which the hydronium ion exceeds the hydroxide ion is acidic.

Since pure water is tasteless and does not affect litmus, and its effect on metals is not at all the same as that of acids, it is the excess H_3O^+ itself that must taste sour, turn blue litmus red, and so on. In pure water the concentrations of the two ions are equal and present only in a minute quantity, and the hydronium ion is no more perceptible than the hydroxide ion. The water molecules themselves we find to be tasteless, and we say they form hydrogen bonds and that they are polar.

The hydronium ion concentration can be increased in several ways. The most common is to add a compound to water which will combine with water in such a way that hydronium ions are produced. All the compounds called acids belong to this group. A typical example is that between hydrogen chloride and water:

$$HCl + H_2O \rightleftarrows H_3O^+ + Cl^-.$$

Hydrogen chloride is called hydrochloric acid, but it does not behave as an acid until it is added to water. Then its hydrogen nucleus migrates to the oxygen of the water molecule, creating an excess of H_3O^+ over OH^-.

Every compound that is called an acid behaves in this manner when it is added to water, and this means that every acid must have at least one hydrogen nucleus, or proton, which it can yield to a water molecule to produce H_3O^+. Many acids have two such protons that they can release, for instance, sulfuric acid (H_2SO_4), and some have three, such as phosphoric acid (H_3PO_4). Obviously the ions, or radicals, that the proton leaves behind will always be negative, or anions: SO_4^{--}, PO_4^{---}.

In the reaction described, if a lot of HCl is added to a fixed amount of water, a lot of H_3O^+ will be formed, and the amount of OH^- will be reduced to a vanishing point in order to keep the product of their concentrations constant. Hydrogen chloride is a strong acid. It ionizes or releases its hydrogen with alacrity. There are weak acids which let their protons go reluctantly, so that only a relatively small number will react with water, and the H_3O^+ concentration will not increase very much. A weak acid is acetic acid, which is found in vinegar. To ionize a fixed amount of it completely—to release every proton—it has to be diluted with large amounts of water, and then, of course, the concentration of H_3O^+ in the total amount of water will still be low. A weak acid is one that ionizes only slightly when it is added to small amounts of water, whereas a strong acid ionizes, or reacts with water, vigorously.

But this property of an acid—of donating hydrogen nuclei, or protons, to form a covalent bond with a water molecule—is also the property of the hydronium ion itself. H_3O^+ donates its extra proton to OH^- in the reversible reaction:

$$H_3O^+ + OH^- \rightleftarrows H_2O + H_2O.$$

Thus we can enlarge our definition of acid to include any compound which will donate a proton to any other compound.

Bases

Suppose now instead of increasing the hydronium ion in pure water, we remove it or suppress it, or in other words we increase the concentration of OH^- at the expense of H_3O^+. Such a solution will have a soapy feel, it will taste bitter, turn red litmus a blue color, and react with an acid to produce a salt. These are exactly the characteristics of a group of compounds called bases. One kind of base is therefore water in which the OH^- outnumbers the H_3O^+ by a significant amount.

The most common method for making a base is by adding to water a compound, such as sodium oxide, Na_2O, that will reduce or remove H_3O^+ and thereby increase OH^-:

$$Na_2^+O^{--} + H_3O^+ \rightleftarrows 2\ Na^+ + OH^- + H_2O.$$

The hydronium ion gives its proton to the sodium oxide molecule, which is thereby rearranged to produce water and hydroxide ion, and sodium ion. The OH^- concentration will be raised, while the H_3O^+ concentration will be reduced. The solution becomes basic.

There are also compounds called alkalis, which already contain the hydroxide ion, and these when added to water simply release their hydroxide ions and make the solution basic that way. Sodium hydroxide, NaOH, known commonly as soda lye, is one such compound:

$$Na^+OH^- \rightleftharpoons Na^+ + OH^-.$$

There are both strong and weak bases.

Neutralization

A strong acid attacks living tissue, causing frightful pain. A strong base or alkali also destroys cells and tissues with a terrible determination, eating into the flesh rapidly. If a strong acid, hydrochloric, and a strong base, soda lye, are added together in a watery solution, the reactions are

$$HCl + H_2O \rightleftharpoons Cl^- + H_3O^+$$
$$NaOH \rightleftharpoons Na^+ + OH^-$$
$$HCl + NaOH \rightleftharpoons Na^+ + Cl^- + H_2O.$$

The two destroyers of flesh neutralize each other, producing ordinary table salt and pure water. If we boil away the water there will remain the white salt, sodium chloride.

In this reaction the hydronium ion produced by the hydrochloric acid and the hydroxide ion added by the sodium hydroxide react instantly to form water in order to keep the concentrations of the two ions at the required constant. If we calculate the exact weights of acid and base to use for a perfect neutralization, the number of H_3O^+ molecules left in the solution will equal the number of OH^-.

The process of neutralizing an acid with a base is one of the most important reactions in chemistry and one of the oldest discovered by the alchemists. It is vital to life. Also, in our modern civilization, half a dozen acids and bases, with half a dozen metals and two or three salts, are the primary materials for that basic industrial activity on which all industry depends.

Let us re-examine the acid-base reaction.

Since an acid neutralizes a base and since an acid by definition donates protons, a base must accept protons. Plainly, the reaction between hydronium and hydroxide ions is itself an acid-base reaction. The H_3O^+ donates its protons to OH^-, which accepts them, both turning into water. The H_3O^+ is the acid and the OH^- is the base, and they neutralize each other.

But the molecules of water that ionize on colliding are also involved with transferring a proton from one to the other. Therefore some few molecules of water are capable of donating protons and must be seen as acid, while some few molecules of water accept protons and must be seen as a base. Water, pure water, is therefore a mixture of neutral molecules of H_2O, of acidic and basic H_2O molecules, of H_3O^+, and of OH^-. All these are maintained in a balanced relationship of concentrations that varies with the temperature.

Thus, according to our definition, every reversible reaction between an acid and a base produces another acid and another base. On either side of the equation sign, therefore, a proton is being transferred. All acid-base reactions are therefore called protolysis, and under this name have been collected a large number of reactions that older acid-base theories could never account for. The definition is still not completely satisfactory and there is a new theory, broader still, but the concept of protolysis has been so eminently useful that at the moment it has general acceptance.

The pH factor

In pure water, or in a perfectly neutralized acid-base solution, the number of hydronium ions and of hydroxide ions is the same. Any increase or decrease of the hydronium ion relative to the hydroxide will turn the solution into an acidic or a basic one. At perfect neutralization each ion's concentration is the fraction 1×10^{-7}, and the product of the concentrations comes to the figure 1×10^{-14}, which is the constant for any kind of water solution at room temperature. No matter what other ions are present, no matter how many are present, $[H_3O^+] \times [OH^-]$ must equal 1×10^{-14}. (To refresh the reader's memory: that last figure is the fraction 1 divided by 10 followed by 13 zeros.)

If the factor $[H_3O^+]$ increases from 1×10^{-7} to 1×10^{-6} then $[OH^-]$ must be reduced to 1×10^{-8}.

In order to make calculations more easily, to get away from handling such fantastically small fractions, the concentration figures are transformed through a logarithmic operation into simple numbers. The fraction 1×10^{-7} becomes the number 7. By convention, the concentration of H_3O^+, rather than that of OH^-, is considered, and we call the concentration of H_3O^+ the *p*H value of a solution. The *p*H of a neutral solution is 7.

The *p*H of a slightly acidic solution would be less than 7, and the *p*H of a strongly acid solution, in which the concentration of H_3O^+ is high, would be 4 or 3. Absolutely pure H_3O^+ would have a *p*H of zero, which, however, is impossible.

A slightly basic solution has a *p*H just over 7. A pure base, in which the H_3O^+ concentration is nil, would have a *p*H of 14, also an impossibility. Thus we can report acidity by a simple number which is derived from the concentration of the hydronium ion. The reader should see that, because of the dynamic equilibrium between water and its ions, it is enough to report the concentration of only one of them. The other can be found from the formula. Why is the *p*H of a strong acid 1, and of a base 13? Because the mathematical conversion that is used reverses the fractions.

Every aqueous solution has a *p*H value. Blood is slightly basic, or alkaline, with a *p*H of 7.3. The body maintains this alkalinity, or basicity, in an unchartably complicated manner through endless reversible reactions in different parts of the body and through different rates of reactions in the glands, and through ionizing compounds called buffers, which allow a certain amount of change in the concentration of acid- and base-forming compounds, without allowing the *p*H to change. The products of all these reactions meet in the bloodstream, and any change in the blood's *p*H factor will at once affect all those reactions, and all those parts of the body that function normally with that *p*H value, until the proper alkalinity is re-established. Even a few decimals of changes in the blood *p*H create violent reactions that may be fatal.

Neutralization means to equalize acid and base, or to produce a pH of 7. The laboratory technique for doing this is called titration. If we have an unknown base in a beaker and slowly drip into it an acid whose strength we know, drop by drop, from a burette which measures the volume of the drops, until the pH in the beaker is 7, then the solution is neutralized, and we say that the titration has been carried to the end point.

The end point is revealed by an indicator, which is a dye that has one color in acid solution and another color in basic solution. As the solution approaches neutrality, each drop of acid fades the dye's alkaline color until just one extra drop will begin to turn the color to its acid tint. At that point we can read on the burette how much acid was added and we can calculate how much base was in the sample by setting up the proper equation and using molecular weights.

Indicators are complex compounds. Many of them combine with water to form hydronium ions and can therefore be considered as very weak acids:

$$H^+Indic^- + H_2O \rightleftarrows H_3O^+ + Indic^-.$$

Only a few drops are added to the solution. The molecule HIndic has one color and the radical Indic$^-$ has another color. A high concentration of H_3O^+ will drive the reaction to the left so that in an acid solution the molecules of HIndic will predominate and will give it its characteristic color, which is red in the case of litmus. When OH^- predominates, they will suppress or take out of solution H_3O^+, and so the reaction will shift to the right and the Indic$^-$ radical will dominate. Litmus will be blue. At the turning point litmus will have a pale pink color. One or two drops more or less of acid will turn the solution red or blue.

Each indicator changes color at a different pH value, so that by using the right indicator we can stop the titration at almost any desired pH value. Countless routine checks in the laboratory are acid-base titrations, and other common indicators besides litmus are phenolphthalein and methyl orange. The color change occurs at pH 7, pH 10, and pH 3, respectively, and each is used in different kinds of neutralization reactions.

Every acid and every base has its own properties, sometimes unique, sometimes not; but each is always recognizably different from other acids and bases, physically as well as chemically. But all acids give up protons and all bases accept protons, and after the exchange the acid becomes a base, capable of reaccepting a proton, while the former base, having acquired a proton, is now an acid, capable of donating a proton. Obviously the relative proton-hunger of molecules will determine how a compound will act, and there are compounds which are useful just because they can act either as proton donors or as proton receivers, depending on the situation. Having a proton to give, at the same time it is able to accept more protons. If it meets acids it reacts as a base and if it meets bases it reacts as an acid.

If we use the periodic chart to guide us, we find that metals form oxides which react with water to form hydroxides, which are bases; we find that nonmetals form acids with hydrogen, and sometimes with hydrogen and oxygen. This generality gives us a clue to acid-base formation in terms of electronegativity. The more electronegative an element—the more nonmetallic—the stronger will be its grip on electrons and the greater will be the probability of its forming compounds with hydrogen in which the hydrogen proton can be easily detached.

Going from left to right in any period of the table, therefore, the hydrogen compounds formed by the elements will be progressively more acidic and less basic. In the transition groups, the potential to be acidic or basic will be about equal, and circumstance will decide which way such a compound reacts. Group IA elements form the strongest bases, while group VIIA elements form the strongest acids. Compounds that act either as acids or as bases are called amphoteric.

It should be pointed out that this theory bears the names of J. N. Brønsted and T. M. Lowry, who developed it independently of each other in 1923, and that it replaced Arrhenius' theory of acids, which introduced pH just as we still use it. Another theory has been proposed by Gilbert Newton Lewis, according to which acidity is defined not by proton transfer but by electron-pair acceptance. An acid is a substance that accepts, or attaches itself to, an exposed electron pair on a base.

The Lewis concept returns us to the framework of electron activity, and it is possible that eventually it will replace the Brønsted-Lowry theory. At the moment a mild confusion reigns among the various textbooks. Many authors are impatient with the hydronium ion and, after explaining the Brønsted-Lowry theory, they return to the use of the hydrogen ion with which they grew up—the Arrhenius theory. Other writers explain all theories but stick doggedly to the Brønsted-Lowry protolysis, though they are baffled by certain aspects of base formation. Still other writers, after carefully explaining the Arrhenius and Brønsted-Lowry theories, develop the Lewis theory as being best. All this means only that each theory is successful in certain ways. The Lewis theory is the broadest, which makes it very attractive, and the only reason it has not been elaborated here is because the Brønsted-Lowry protolysis concept has the widest acceptance and has also introduced a vocabulary that seems to be very useful.

Salts

Whatever the theory, the reactions themselves are, of course, the same, and a base and an acid will always react to produce water and a salt, the salt consisting of the metallic radical of the base and of the nonmetallic radical of the acid. The generic word "salt" refers to all compounds which consist of metallic and non-metallic ions, or radicals, whether or not they are produced by acid-base reaction. But here too the relative electronegativity of the ions will produce a balanced rather than a rigid web of attractions and repulsions. The salt obtained from a strong acid and a weak base will, when it is dissolved in water, create an acid condition because it will suppress the hydroxide ion. The pH will be low. Salts from equally strong acids and bases will produce neutral solutions with water. For instance, sodium and chlorine are equal in their ionization potential, that is, Na^0 forms its positive ion Na^+ with the same vigor that Cl^0 forms its negative ion Cl^-. Their electronegativities are opposite but on the same order of magnitude. The salt that results is therefore an evenly balanced affair between the two ions, and the water solution is neutral.

The study of salts is probably as old as man; certainly it is much older than the study of acids and bases, for the simple reason that active acids and bases do not exist in the crust of the earth, whereas salts form a large part of it. There are hundreds of salts, and many elements are obtained from their salts. The commonest salts are those of strong acids such as hydrochloric, sulfuric, and nitric. The oxides of metals are saltlike in character but are not called salts because they do not derive, theoretically, from acid-base reactions. Sulfides, carbonates, silicates are salts associated with the radicals of weak acids.

The points to remember are that all salts ionize, that when they are in the solid state they form a pattern of positive and negative ions rather than a pattern of molecules, that in the liquid state the ions move about freely and so again we cannot speak of an actual molecule of salt; and that only in the gaseous state do we have a cation and an anion stuck together into what may be called a single molecule.

Not only do salts form a large part of the earth's crust, but their extraction, or preparation, is an important part of modern industry. There are various methods for making salts, one of the most ancient being the evaporation of water from a solution of the desired salt. From salts are manufactured both acids and bases, and a large variety of other kinds of compounds, as well as many elements.

This brief reference to a huge group of compounds allows us to digress slightly from our development of chemical information to examine the concept of mixtures and solutions, without which it would be impossible to relate any theory of molecules and atomic structure to ordinary experience.

19

The Significance of Solutions

Molecular freedom in chemical change

A chemical reaction cannot begin unless the reactants are in contact at activation levels of energy, and the reaction cannot proceed if the products clog or prevent fresh collisions from taking place. Molecular freedom of at least one reactant to move about and mingle with the other is a prerequisite, and therefore the study of mixtures is an important field in chemistry.

All mixtures consist of any number of different compounds and elements in any proportion and not bonded in any way to one another, so that the ingredients can be separated by physical manipulations. Some of the more common methods are filtration of a solid from a liquid, distillation of one liquid from another, centrifugal action on two liquids or on a liquid and a solid, evaporation of a liquid to leave behind a solid, and diffusion of gases through finely porous solids that hold up the heavier gases.

Mixtures are of two types, by definition, heterogeneous and homogeneous. Heterogeneous mixtures are those in which the constituents are at least large enough to be microscopic in size, such as muddy water, different-colored clays and sands, sugar and salt in a bowl, soot and air, milk, concrete. Homogeneous mixtures consist of components in uniform molecular or atomic dispersion such that any sample contains the same proportion of components as a result of their kinetic energies, which disperse them uniformly throughout the container. Tea is a homogeneous mixture of water and tea compounds; if sugar is put in a cup of tea, for a while a heterogeneous mixture exists in the cup but after the sugar melts there is again a homogeneous mixture. Every cup of tea can have a different proportion of sugar in it; there is no fixed ratio to the components of either a homogeneous or a heterogeneous mixture.

Homogeneous mixtures are called solutions. The component in greater bulk is called the solvent, and the components dissolved in it are called solutes. Solutions exist in all three states of matter. Liquids and solids can be vaporized to form gaseous solutions, such as ordinary air, in which water, oxygen, carbon dioxide, nitrogen, and other elements are uniformly mixed by their own molecular movement.

Liquid solutions consist of liquids that dissolve in liquids, such as alcohol and water, solids that dissolve in liquids, such as sugar and water, and gases that dissolve in liquids, such as carbon dioxide and water.

Some solids can be considered to act as solvents for certain gases and liquids and for other solids; some metals which melt together can be considered solid solutions on hardening.

The criterion for solution is simply the uniform distribution of components at the molecular level without chemical bonds, and to produce proper solutions for a reaction is one of the problems in industrial chemistry. To turn everything into a gas—the ideal solution—is impossible because many compounds break down before they boil, and because many compounds boil only at very high temperatures, which would make the process too expensive, and because a reaction between gases must be carefully controlled to avoid explosion. The reaction between a gas and a liquid can be better controlled. The gas is bubbled through a liquid in towers designed so that the liquid trickles downward over baffles which allow the rising gas to be in contact with it long enough to react. Liquid solutions poured into one another are perhaps the most common way of bringing reactant molecules together, though almost as common is dissolving of a solid in a liquid reagent.

The forces of attraction between liquid molecules of the same compound, and also between solid molecules of the same compound, are nowhere near as powerful as the forces that mesh the valence electrons between atoms within the molecule. The former are considered to be the result of attractions between any nuclei and all electrons of the nearby nuclei, and were first measured in 1880 by the Dutch physicist Johannes Diderik van der Waals. Different kinds of molecules attract one another with unpredictable strength, but obviously some compounds mix well, while others mix not at all. Oil and water repel, alcohol and water are compatible in any amounts.

There is only one generality—and it is not a law at all—which in its broadest sense states that each type of system tends to attract similar systems, electrical charges excluded. Covalently bonded molecules will therefore mix with or dissolve in covalently bonded solvents better than in electrovalently bonded compounds, while the latter will dissolve other ionic compounds to an unlimited extent. One can even say that polar liquids will dissolve most other polar substances better than they will nonpolar substances. Water, because its molecules are covalent, ionized, polar, and capable of forming hydrogen bonds, is predictably a good solvent for these four types of compounds.

Each molecule of sugar in the crystalline white solid attracts to itself all the molecules near it with a weak van der Waals force. If a spoonful of sugar is put in a glass of water, water molecules press upon the faces of a sugar crystal and penetrate any interstices they find and weaken with their own attractive forces the van der Waals forces in the crystal. The water presses deeper into the crystal and pries the sugar molecules loose. They swim out among the water molecules as free agents. If the water is heated it bores more persistently, and the sugar molecules, themselves vibrating with greater kinetic energy and standing farther apart, allow more water molecules to press in. The sugar melts more quickly. Stirring also hastens the dissolving process by carrying away the sugar molecules in currents of water.

Most compounds dissolve with the release of a fixed amount of energy, called the heat of solution, which is different for each compound and is not the same as the heat of fusion, and some draw heat into themselves as they dissolve, cooling the solution.

In any case this heat is also an identifying property of many compounds and is listed in the handbook of chemistry.

There is a limit to the number of sugar molecules that a beaker of water will tolerate in itself. One can think of dissolved sugar as a cloud of gas in the water, filling all the space between the water molecules uniformly until there is no room there for more sugar. If the water is heated it expands, the space between its molecules increases, and more sugar molecules can be accommodated. More will dissolve and fill that available space. Thus at every temperature there is a maximum of sugar that can be dissolved in a fixed amount of water, as though there existed a limit of tolerance between two kinds of molecules. The higher the temperature, the greater the tolerance. With a few exceptions this rule holds for all kinds of solutions except for those of a gas in a liquid or a solid. In such a solution, if the temperature of the liquid or the solid is raised, the dissolved gas will escape more easily than if the temperature is lowered, because of its kinetic energy.

A solution that contains all the solute that it can dissolve is called saturated. If more solid is added to a saturated solution, it cannot dissolve, but a dynamic relationship is set up between the undissolved solid and the dissolved molecules. Solid molecules of high energy will force themselves into the water, but, since they exceed the tolerance, the same number of molecules of weak energy settle out of the solution onto the solid, as though they had been squeezed out. A crystal of a compound suspended in its own saturated solution is constantly bombarded by dissolved molecules, and an incessant interchange goes on at the crystal's surface. If the temperature and solubility are raised, more of the crystal melts; if the temperature and solubility are lowered the crystal grows, with dissolved molecules coming out of solution to deposit on it.

The concentration of a solution is reported as the weight of solute dissolved in a certain weight of solute plus solvent. A strong or concentrated solution contains a great deal of solute. If the liquid of a weak solution is evaporated, the concentration of the solution increases steadily. When enough water has been removed from a sugar solution that is not saturated, a concentration is reached when it will be saturated, and further evaporation will force some of the sugar out of solution. If evaporation is carried out carefully, many solutions will become supersaturated, which means that more than the permissible amount of solute is actually in solution. When this condition is disturbed by the addition of a crystal, or when it is shaken, the excess solute at once deposits out of the solution.

The concentration of a saturated solution is different for each compound. If a mixture of solids is dissolved in an appropriate liquid and the liquid slowly evaporated, each solid will begin to crystallize out at its own point of saturation. This method can be used to separate the compounds in a mixture of solids, or to purify any compound of its impurities.

Liquids which are soluble in each other are called miscible, and for some liquids—for instance, alcohol and water—there is no limit to the proportions that can be used. Other miscible liquids have limited solubilities. The liquid present in larger bulk is usually called the solvent. If a solution of two liquids is cooled sufficiently, one of them will freeze before the other, and this may be the best way of separating them.

Solutions of solids and liquids in solids are special cases which seldom occur in chemistry, but the kinetic molecular theory applies the same reasoning to their properties.

The handbook of chemistry always lists as an identifying property of an element or compound how many grams of it will dissolve in 100 grams of water at a certain temperature, and this figure is called the solubility of the compound. Solubilities are also given for alcohol and ether and other useful solvents. Instead of a figure, the solubility is often reported as being slight, or moderate.

Standard solutions

Since it can be said that solutions are the best media for chemical change, the preparation and handling of solutions is one of the prime problems in industry, where the cheapest methods for the largest yield of product have to be sought. However, huge quantities of liquid reagents, especially acids and bases, also must be prepared for laboratory work in concentrations that are known with fine precision. The amounts of a reagent that are required for laboratory work are always small and often minute, and it is far simpler to measure out a substantial volume of a solution which contains the minute quantity desired, than it is to measure out that tiny quantity directly on the balances. Glass pipettes and burettes are graduated to handle as little as $\frac{1}{100}$ of a cubic centimeter of liquid, which is hardly one drop.

The mole of a compound is that amount of it which weighs as many grams as the figure of the compound's molecular weight. A mole of HCl is 36.5 *grams*. If 1 mole of a compound is dissolved in 1000 cubic centimeters (1 liter) of water, the solution is called 1 molar. A 1-molar solution of hydrochloric acid consists of 36.5 grams of acid in 1 liter of water. Thus each cubic centimeter contains 0.0365 grams of acid, and each tenth of a cubic centimeter contains 0.00365 grams of acid, a quantity that it is impossible to weigh out. By measuring the volumes of a molar solution that have to be used to neutralize an unknown sample of a base, we can calculate the amount of base in the sample. A molar solution is defined as one that contains 1 mole of solute in 1000 grams of solvent.

There are several other types of standardized solutions. The word "normal" is used to indicate the availability of a single proton in acid-base reactions. Thus a normal solution of HCl is the same as a molar solution of that acid. But a normal solution of H_2SO_4 contains only 49 grams of sulfuric acid in 1000 cubic centimeters of water, instead of 98 grams, which is 1 mole of the acid, because it has two protons to donate.

Without such standardized solutions, a chemist's work would never get done, and that statement suggests correctly that it is the liquid phase in which most reactions take place.

We have kept till the end one of the most important aspects of solutions—the difference between covalent and electrovalent, or ionic, compounds when they are dissolved in water.

It was found that any compound dissolved in water lowers the water's freezing point and raises its boiling point. If 1 mole of a covalent solute is dissolved in 1000 grams of pure water, the freezing point is lowered to $-1.86°$ C. and the boiling point is raised to $100.52°$ C. For each mole added, the change increases by the same amount. The kinetic molecular theory explains this by focusing upon the vapor pressure of water and thus upon the actual number of solute and solvent molecules present.

If we now dissolve 1 mole of an electrovalent compound in pure water, the freezing point is lowered by more than $1.86°$ C. This can be explained by the kinetic molecular theory, reaffirming that since the determining factor in this phenomenon is the actual number of dissolved particles, not their weight, the electrovalent compound must break up into its ions, thus adding many more particles to the solution than if it were a covalent compound.

All acids, bases, salts, metallic oxides break apart into their positive and negative ions when they dissolve in water and in a few other solvents. The ionization may be slight with some of these electrovalent compounds when there is not enough water present, but dilution will always bring about 100 per cent ionization. Strong acids and bases and their salts ionize a great deal even in small amounts of water—that is, at high concentration.

The resulting solution is called an electrolyte because it conducts electricity, and we examine electrolytes and the conductivity of metals in the following chapter.

20

Electric Current and Chemical Change

Metallic bonding

WE know that the force with which an element becomes an ion can be determined by measuring the force with which it clings to its valence electrons. This force, called electronegativity, is discussed in Chapter 16. If two atoms with similar electronegativities combine, they will have to share electrons in a covalent bond. If two atoms differ somewhat in their electronegativities, then the stronger will hold the shared pair nearer to its nucleus and the molecule will be polar—more negative at that end. But if the electronegativities of the combining atoms are widely different, then the bonding electrons will be actually transferred from the less to the more electronegative atom, and the atoms become ions which can exist separately but which normally cling together because of their opposite charges.

The explanation for this shifting of valence electrons is that certain electron configurations are more stable than others, and the most stable is that of the inert gases, all of which have 8 electrons in their outermost shell. Thus, all reacting atoms either shed or gain or share electrons in order to count 8 in their outermost shells, because the total free energy in the ion configuration of an element is less than the total free energy in the atom configuration of that element.

Since strong metals have only a few electrons in their valence shells and have low electronegativities, they cannot form covalent bonds among themselves, nor can they form electrovalent bonds with one another. Eight atoms of sodium would have to adjust themselves in such a way that each one's single valence electron would weave itself around all 8 atoms. As for electrovalent bonds, one sodium atom would have the impossible task of taking in 7 electrons and becoming an anion with a charge of —7. Sodium barely hangs on to its single electron and will never accept more. Therefore a metal cannot combine chemically with another metal by shifting electrons between atoms to form molecules.

Yet some kind of bonding must be discovered to explain the fact that transition-metal atoms form the strongest solids we know, with several special properties that distinguish them from all other kinds of solids: they conduct heat, they conduct electricity, they are magnetic, and they are malleable and ductile. Not all metals have all these properties, but nonmetals have none of them, except under certain conditions.

We know that a current of electricity is a flow of electrons along a conductor. Does a copper atom conduct electricity or must copper atoms be linked together into a solid before electrons will flow? Does each copper atom in a wire pass an electron along? If copper and other metals do this, why won't compounds such as sugar, or elements such as sulfur?

Obviously the explanation of electrical conductivity must jibe with the explanation of how metal atoms link to form the solid state so that they can be bent, pounded thin, drawn into wire, hardened, softened, sharpened, tooled, stamped, cut, whereas no crystal of a nonmetallic element and no solid of a covalent or electrovalent compound will take a fraction of such harsh treatment.

If we imagine a metal wire with the atoms cheek by jowl, since it does not break or crack when being bent, we see that, the atoms on the outside of the bend being pulled apart and the atoms on the inside of the bend being squeezed together, the inside ones must be forced up into the open spaces left by the outside atoms. If we pound a sheet with a hammer it flattens out without cracking. Again we imagine the outside atoms being squeezed down by each blow and the inner atoms being separated sideways to accommodate the downward pressure. If we imagine a wire being pulled through a hole smaller than the wire, we can imagine a roll of atoms at the entrance to the hole being squeezed along the wire as it is pulled through.

The phenomenon cannot be explained by the kind of intermolecular forces that we have identified in solids as van der Waals forces. In the crystal, each ion or atom or molecule remains fixed into the lattice of nuclei and electrons, so that a blow will shatter the crystal, or shear it, while squeezing will turn a crystal to powder. Nonmetallic atoms and molecules cannot be made to slide about without breaking their forces of cohesion.

Also, when a metal is heated it gradually softens, quite unlike a crystal, which may expand but which retains its specific structure up to the melting point. But metals have a sharp melting point too and give up their heat of fusion when they become liquid, and that is why they are true solids and not just congealed liquids, like glass.

True, not all the elements on the left-hand side of the periodic table that we call metals can be rolled and flattened, but even those that cannot be are not at all like crystals. It is in the incompleted *d* and *f* subshells inside the outermost *s* orbitals, of the transition metals, that the metallic properties so familiar to us originate.

The explanation in its most general form is that in a solid transition metal each atom has been pulled so close to all the atoms immediately surrounding it that all the valence shells have been penetrated and the shells just inside the valence electrons are thus in very close contact. Hence, each atom's cloud of electrons is as close to a neighboring nucleus as it would be in a chemical bonding, while in the ordinary solid phase of nonmetals no penetration whatever of the electron clouds occurs.

The 1, or 2, or 3 valence electrons that a metal may have can be considered, in the solid state, to be held with very little attraction by any of the neighboring nuclei, which are more intensely associated with the inner electron shells of their neighbors. The valence electrons are also repelled by the inner shells of neighboring atoms. Thus the valence electrons lose all sense of belonging, as it were, to any specific nucleus, and so every atom can count its valence electrons disposed of without actually losing them.

Between the tightly packed atoms there are always spaces into which the valence electrons can move, and there the pressure of positive and negative forces is such that if an electron is lost from such a space another moves into it from somewhere else. These places have associated with them definite energy levels, and quantum mechanics charts their relationship to the electron subshells that are in contact in the solid metal. In any piece of ordinary metal, of course, the total number of electrons equals the total number of protons, and not only is the whole piece electrically neutral but each atom in it is neutral. And yet the valence electrons can drift from one low-energy-level spot to another without disturbing the rest of the electrons. In fact these spaces can be considered as relatively positive areas when not occupied.

These valence electrons in a solid metal may be pictured as a cloud of free-to-move electrons, free to flow about in the empty spaces between the atoms. They rest quietly in the hollows of low energy where the multiple fields of attraction and repulsion all around them are most in balance. A minor variation in the balance of energies creates waves of changing kinetic energy in the vibration of the atoms, and the valence electrons flow a little this way or that, always keeping themselves distributed against one another's repulsion and the repulsion of the electrons tied strongly into their inner shells.

We can imagine how the valence electrons, being free of any specific atom, allow the rows of atoms to be bent and displaced. Atoms can be made to slide about under physical pressures, always staying linked to their neighbors, while the valence electrons drift from one neutralizing point near a grouping to the next. There is always space for only a fixed number of electrons.

Electron flow in metals

Suppose outside electrons are forced into a metal strip. The effect of these outsiders upon the first loosely held valence electrons will be to set them flowing away from the pressure, away from the source of the new electrons. The electrons will be knocked along from space to space between the atoms. A current of electrons will flow along the metal strip as long as the pressure of new electrons lasts. The electrons at the other end of the metal strip will be knocked clean out of the strip and will leap in whatever direction local fields of magnetism and electricity indicate. If the pressure of new electrons stops, the flow of electrons along the wire stops at once.

This picture, based on a mixture of theories, explains many facts about electric currents and has proved to be flexible enough to accommodate data obtained by research into the electrical properties of alloys, of semiconductors, which are impure crystals, of nonconductors, and even of the odd magnetic property of some elements in the solid state, such as oxygen. In fact, this picture seems capable of expansion and refinement to include even broader fields of atomic relationship. It is part of the whole concept of electron configurations, orbitals, quantum numbers, levels of energy, dynamic equilibrium. We do not have to abandon any of the general principles of positive and negative charges or bring in any new principles of atomic structure. Metallic bonding is simply another form of bonding through valence electrons with this difference: no chemical transformation takes place when metal atoms settle into a solid state and establish their metallic bonds— that is, no chemical transformation in the conventional sense of the phrase. But obviously when a metal is melted into a liquid it will lose its drifting cloud of electrons and hence its property of conducting electricity and, though we will not investigate it here, its magnetic properties. The point is that the nature of the bonding between metal atoms is nothing like the nature of the van der Waals forces binding molecules into a crystal.

We can now state that the forces which create the actual structure of an atom and establish its properties, and which shape the configuration of molecules and establish their properties, are the same forces that establish a metal's electrical properties. We have related the actual flow of electricity along a metal conductor to the modern concept of chemical change, and we will now examine the actual mechanism that transforms chemical change into an electrical current, and vice versa. We begin with the concept of what happens when a compound breaks up into ions and radicals.

Electron flow through electrolytes

The bond between sodium and chlorine in Na^+Cl^- is produced by the attraction between negative and positive charges. If we imagine either molten salt or solid salt, should we imagine closely knit particles, molecules, of NaCl associated with one another by weak van der Waals forces? Or should we imagine that any Na^+ is equally attracted to any Cl^- in its vicinity? That is, does each Na^+ cling exclusively to the same Cl^- at all times? Or does each Na^+ cling to whatever Cl^- happens to be nearest? Since the bond between any two ions is the same strength as the bond between each ion and any other of opposite sign, we can imagine the ions in a crystal arranged in such a way that every positive ion is surrounded by negative ions and each negative ion is surrounded by positive ions. And indeed this is the way ionic compounds form crystals. The bonds between any one Na^+ and all the Cl^- to which it is attached around it are exactly the same. There is, in other words, no perceptible molecule of NaCl. The whole crystal, however large, might be considered the molecule.

If the crystal is melted, the sodium and chlorine ions acquire the freedom to move about. Obviously, no sodium ion will cluster with other sodium ions; each will surround itself with chlorine ions. Throughout the liquid salt—or, to use the proper name, fused salt—there is a uniform alteration from positive to negative ion in all directions. That is, one always finds some one Cl^- balancing some one Na^+, but as the liquid is stirred or poured the ions slide about independently, and not as a Na+Cl− particle.

We can safely say that it is possible only in the gaseous state for a single pair of ions to exist as a molecule in the conventional sense of that word.

Suppose we dissolve the salt in water. The water molecules pry apart not Na+Cl− particles, but separate ions of Na+ and Cl−. The solution will consist of H_2O, H_3O^+, OH^-, Na+Cl−, Na+, Cl−; there will be polar water molecules stuck end to end, and water molecules bound with hydrogen bonds to other water molecules, and all these ions and nonions will be in a state of dynamic equilibrium, constantly shifting their positions but maintaining whatever concentrations the temperature and the strength of the solution set.

Pure water conducts electricity to a very slight extent. Water with any electrovalent compound dissolved in it conducts electricity quite well; therefore any ionic compound is also called an electrolyte.

The electric battery

When a strip of metal is put into pure water, some of the metal atoms will be pried off by the water, and this process allows a lower energy state to be reached by the metal atoms if they abandon their valence electrons in the strip and swim off through the water as ions. If the metal is zinc, the following equation can be written:

$$Zn^0 \rightleftarrows Zn^{++} + 2e^-.$$

The $2e^-$ having been left on the strip by the dissolved Zn^{++}, the strip now has an extra 2 electrons. The strip is now negatively charged, while the water is now positively charged by unbalanced protons of the Zn^{++} ions. Only a small number of zinc ions can go into solution before the strip has enough abandoned electrons on it to prevent any more free electrons from joining them, and thus to prevent the escape of any more positive ions from the strip. A dynamic equilibrium will be established between the zinc atoms in the strip and the zinc cations in the water. A constant exchange will be going on, as some of the zinc cations on hitting the strip will be caught and will reaccept electrons to become neutral atoms again.

The equilibrium depends on the relative forces on either side of the equation. If we remove the electrons from the zinc strip, then more zinc atoms will go into solution as ions, leaving behind their valence electrons. If we somehow remove the zinc ions from the solution, again more zinc atoms will be able to leave the strip and become cations. In each case a point will be reached when the build-up of either electrons or zinc ions will prevent further movement. But if both electrons and ions are removed steadily, the strip will continue to dissolve into the pure water, until the whole of it has been transformed into zinc ions and electrons.

Many metals behave in the same way. If we put a copper strip into pure water, some copper atoms will go into solution as copper ions, leaving behind their valence electrons to give the copper strip a negative charge.

Now the pressure to ionize in the zinc strip is greater than the pressure to ionize in the copper strip. That is, the zinc atoms become ions more vigorously than the copper ions, piling up a larger number of electrons on the zinc strip before equilibrium is reached. Thus we can imagine a device which will allow the higher pressure of electrons on the zinc strip to flow away toward the lower pressure in the copper strip.

But first imagine the zinc strip in a solution that contains copper ions, because an ionic compound of copper has been dissolved in it, for example, copper sulfate, $Cu^{++}(SO_4)^{--}$. The "eagerness" of any zinc atom, Zn^0, to give up its electrons is greater than the "eagerness" of any copper atom, Cu^0, to do so. That is to say, copper is more electronegative than zinc. Each time a Cu^{++} hits the Zn^0, a Zn^0 atom will gladly force its valence electrons on the Cu^{++}, which will accept them, and thus become a Cu^0. The Zn^0 goes into solution as a zinc ion, while the copper ion plates out on the zinc strip as Cu^0.

Copper ions in solution are blue. After the zinc strip has been in the copper solution for a while the blue color will have faded, and eventually it will disappear, while the zinc becomes coated with reddish-black copper metal.

We can write the following equations for what happens:

$$Zn^0 \rightleftarrows Zn^{++} + 2e^-$$
$$Cu^{++} + 2e^- \rightleftarrows Cu^0$$
$$\overline{Zn^0 + Cu^{++} \rightleftarrows Zn^{++} + Cu^0.}$$

The zinc was oxidized, while the copper was reduced, according to our definition of oxidation-reduction, or redox, reactions.

Now imagine a zinc strip in a solution of $Zn^{++}(SO_4)^{--}$, zinc sulfate, and a copper strip in a solution of copper sulfate, $Cu^{++}(SO_4)^{--}$, the two solutions being separated by a porous membrane and the two strips connected by conducting wire.

The membrane can be of different kinds of material, including unglazed earthenware, as long as it acts like a revolving door for molecules. Ordinary random bouncing and flying about do not allow the molecules or ions to pass through the intricate, tortuous passageways of the membrane. They bounce back from it. But if there is a steady pressure applied to them, or a steady pull, they will persist in hitting the membrane more and more often in the right direction, and gradually they will migrate through it. Such membranes are a common piece of laboratory equipment.

In the zinc metal there will be a pressure of electrons, and in the copper metal there will be a smaller pressure of electrons than in the zinc. Since the two strips are joined by a wire, the higher electron pressure in the zinc will force electrons along the wire into the copper, leaving the zinc somewhat positively charged and giving the copper a negative charge.

The Zn^{++} in solution will be attracted toward the copper strip, and so will the Cu^{++}. These ions take up the extra electrons that have packed into the copper strip from the zinc strip and thus allow more electrons to flow from zinc to copper and more zinc atoms to become Zn^{++}.

In the meantime, since all ions under pressure find their way through the porous wall, all the SO_4^{--} migrates toward the zinc.

This continuous shift of ions in the solution, of zinc atoms becoming positive ions, of copper ions becoming neutral atoms, produces a continuous flow of electrons along the wire from the zinc to the copper. If a light bulb is connected to the wire it will light from resistance to the flow of the electrons through it. What we have in this simple set-up of metal strips and dissolved salts is a system for making electricity—a voltaic cell named after the Italian physicist Conte Alessandro Volta, who discovered the principle in the early nineteenth century. It is the same principle that is elaborated into the different kinds of electric batteries that we use today for so many purposes.

The difference between chemistry and electricity has vanished, or rather we have discovered that the potential for chemical transformation in the atoms of the elements can be converted into an electrical current.

It must be remembered that electronegativity is purely a relative concept. Fluorine was assigned an electronegativity of 4, but there is no basic unit of electronegative measure. The potential for seizing or donating electrons is relative. A voltaic cell can be constructed with copper and silver electrodes, or poles, in which the copper is the less electronegative of the two metals and plays the role zinc did in the previous illustration. The electrons would then flow from the copper electrode to the silver electrode.

One can say that copper is more actively metallic, or is more easily oxidized, or has a higher oxidation potential when in contact with its own ion, than silver, but copper is less active than zinc. Thus the elements can be listed in order of decreasing activity, or oxidation potential, when in contact with their ions. In such a list, called the electromotive series of the elements, each element would be more electronegative than all the elements above it and would accept electrons from any of those elements.

The standard which has been set arbitrarily is the force with which hydrogen becomes an ion—becomes oxidized, or loses its electron—and this is given a zero value. (In this connection hydrogen acts like a metal.) Actually it is the condition of equilibrium that exists between hydrogen gas constantly bubbling through a solution containing hydrogen ion, but the apparatus with which this is accomplished, and how it is connected to the apparatus which contains metals in contact with their own ions, and how the relative pressure of electrons, and the electromotive force developed between the two, is measured, need not be detailed here. With hydrogen set at o, any metal has a positive or negative potential against hydrogen. Any metal used in a voltaic-cell arrangement will force all metals listed below it to accept electrons, but will give up its electrons to all metals above it in the electromotive series.

Most batteries are built to deliver rated electromotive force between their electrodes, or poles. This requires an arrangement of elements and ions such that drawing off the current of electrons does not clog the system with the products of the reaction until the reactors have been used up. Batteries intended for heavy duty consist of many cells linked together in such a way and constructed from such materials that, when a flow of electrons is pressed into them, the reactions are reversed and the original components are reproduced within the cells. Thus the battery is recharged. In some batteries the electrodes are neutral, not metallic, and the electromotive force is provided by two solutions consisting of a mixture of ions, one of which has a higher oxidation potential than the other.

In any case, in every kind of battery, redox reactions are going on at both poles and, because of the difference in the ease of oxidation, a pressure of electrons builds up on one of the poles. Electrons can be drawn off at this pole if, at the same time, the circuit is closed by connecting the other pole to the system using the flow.

Electrolysis

Now imagine two conductors leading from the poles of a battery to a liquid electrovalent compound which has ionized so that its cations and anions are moving about freely. The chemical reaction in the battery between its poles and ions presses electrons into the negative electrode, or cathode, immersed in the liquid ionic compound, or electrolyte. Relative to this pressure of electrons on the cathode there is only slight pressure or, in a sense, a dearth of electron pressure at the anode. In the vicinity of the cathode all the positively charged ions accept electrons piled up out of the battery, and become neutralized to atoms. From all parts of the liquid there will be a general movement of positive ions, or cations, toward the cathode, where they become reduced to neutral atoms. At the same time, anions will be able to give up their electrons to the anode, where there is a lack of electrons as a result of the electromotive force within the battery's cells. All negative ions will move toward the anode, and these become neutral atoms.

If the liquid is melted or fused salt, it consists of freely moving Na^+ and Cl^-. The Na^+ will go to the cathode and accept an electron, becoming Na^0 metal, while Cl^- will go to the anode and give up its electron to become Cl^0 atoms, which at once combine into chlorine molecules, Cl_2, a gas. The fused salt has been electrolyzed. The pressure of electrons from the battery, the electrical current, has reduced the sodium ion to the metal and has oxidized the chloride ion to gas. In the process, electrons have been removed from the cathode and other electrons have been supplied to the anode, so that the current has flowed back into the battery. A large amount of chlorine gas and sodium metal is manufactured by the electrolysis of salt in specially built apparatus that will allow the safe collection of sodium metal and chlorine gas.

Suppose we dissolve the salt in water. Now we have H_2O molecules, H_3O^+, OH^-, Na^+, Cl^- all uniformly distributed in the solution. Both the Na^+ and the H_3O^+ will move to the cathode and both the OH^- and the Cl^- will move to the anode. Such a complicated sequence of oxidation-reduction reactions occurs that the true mechanism has not yet been clearly discerned. The fact is that the end products are hydrogen gas instead of sodium, and chlorine gas.

If we electrolyze a weak water solution of an acid the products will be hydrogen gas and oxygen gas, through a complex series of reactions, at the cathode and anode.

If any kind of metallic salt is electrolyzed, the pure metal will plate out on the cathode, and if the cathode is an object like a copper spoon and the solution contains ions of silver, the silver will be deposited on the spoon, coating it. Electroplating is a simple application of electrolysis. A solution of chromium ions will plate an iron object connected to the source of the current as the cathode.

Thus, a solution of any electrolyte will conduct an electric current; that is, some ions will pick up electrons at the cathode while other ions will deliver electrons to the anode, allowing the electrochemical or oxidation-reduction reaction in the battery to proceed.

It would now be logical to begin an examination of the elements, individually and as members of the groups in the periodic table—their natural occurrence, methods of preparation, physical properties, the chemical reactions they undergo, the compounds they form, their usefulness, their industrial importance, their resemblance to other members of their group, the things about them which are unique or historically interesting. Especially, the reactions around which engineering techniques have been developed could be carefully studied. But all that information can be found in any general textbook on chemistry in details we could not reproduce here, as well as in several books written for the layman on the history of chemical industries. In this introduction we will not expand on specific knowledge of the elements and their compounds, except in the case of one element: carbon.

If we look again at the grouping of the elements in the periodic table and examine the compounds formed by them, we find that elements in the same group form similar compounds. The compounds of lithium are much like those of sodium and potassium and cesium—the alkali metals. The rare-earth elements form compounds so much alike that it is difficult to separate them from one another in a mixture. Group VIIA elements, the halogens fluorine, chlorine, bromine, and iodine, form compounds that are very much alike. A compound's properties, however, are not just the sum of the properties of its elements, so that it can be examined as a derivative of any one of the elements in it, but in many cases there is a dominant element in a compound. For instance, the hydrogen in acids plays a role that hydrogen in no other kind of compound can play, and thus all acids have in common their ability to donate protons. At the same time, the different acids are best studied as compounds formed by different elements, such as sulfur or phosphorus or chlorine, and we find the acids formed by these elements listed as compounds of sulfur or phosphorus or chlorine.

Hence, the periodic table creates an orderly study of the thousand or so inorganic compounds that we know, in a way that is not merely practical but also theoretical. A compound consisting of three elements ABC is not studied under all those three elements but only under one of them, say B, the one whose neighbors in its group also form similar compounds with A and C. The significance of the compound because of its prevalence in nature, or because it is very rare, or because it is important to industry or to man's survival, or because of some historical usage, may cause it to be placed with one element rather than another. Pyrite, FeS_2, is always studied with iron because it is a source of iron in metallurgy, although it is also used as a source of sulfuric acid.

In the case of carbon the method of linkage between carbon atoms is unique, and thus it is correct to study all carbon compounds as derivatives of carbon linkage, rather than as oxides or acids or salts.

21

The Infinite Variety of Organic Matter

The carbon atom

THE ancients, the alchemists, the early-nineteenth-century scientists were convinced that the substances found in association with any living organism were created by the mysterious process of organic existence and, as a result, contained a special life force. Ordinary chemicals, it was believed, could be prepared from basic ingredients, but only the processes of life itself could prepare the chemicals necessary for life. Expressed in modern terminology, this view postulates two kinds of compounds: those that could be synthesized, or put together, from elements in the laboratory; and those that only living organisms could put together from the elements. Not only was it impossible to make leaf-stuff and flesh-stuff in a test tube, but even the much simpler compounds found in the leaves and in flesh could not be synthesized without the influence of life forces. For this reason, all such compounds, even the simplest, were called organic compounds. Flesh and bones, leaves and roots, all manner of substances in living creatures and plants, were constituted of organic compounds. Man could only destroy, break down, analyze, separate into their elements these compounds, but he could never put them together again into flesh and leaves and the compounds associated with the life process.

Inorganic compounds comprised the rocks, the earths, the waters and air, the mineral kingdom, the inanimate bulk matter of the universe, consisting of compounds that man could manipulate, break up and put together in the laboratory, endlessly.

Then in 1828, one of those rare dates remembered because a single and rather modest discovery became a revolutionary turning point in knowledge, the German chemist Friedrich Wöhler accidentally synthesized urea from inorganic compounds. Urea is found only in the body wastes and is indubitably organic in character. Wöhler had proved not only that organic compounds could be synthesized outside the living organism but also that the raw stuff for the synthesis could be inorganic. Therefore there was no life force, no intangible power in the living organism, that pulled the inorganic elements into structures that could not be imitated by scientists in the laboratory, using ordinary chemical apparatus and inorganic materials.

Following Wöhler's discovery every organic substance was attacked in every conceivable way to break it down and analyze its composition and to determine its formula. When the formula was deduced, the researchers could set about trying to synthesize the compound. The emotional effect on the scientific and lay public is difficult to evaluate today, but it is not too much of an exaggeration to say that without this development in chemistry Darwin's theory of evolution, pieced together out of various theories, could not have been brought to its full illuminating power, less than a generation later.

In the search for methods of synthesizing known organic compounds, chemists discovered that they could create completely new compounds not found in nature anywhere. The organic chemist became the most specialized scientist of all, and at the same time the most adventurous, and by the middle of the nineteenth century organic chemistry was the first science to pay for research, and pay handsomely. Men who discovered ways to synthesize rare or expensive natural stuff became rich. The modern chemical industry was born with aniline dyes, and today plastics, synthetic fibers, pharmaceuticals, insecticides, continue to make organic chemistry a rewarding field for research.

Around a million organic compounds are known, and a large percentage of these never existed before man created them. Obviously a new definition of "organic" must be attempted.

There is one element that occurs in every natural and synthetic organic compound, and that is carbon. Therefore the whole field is now called the chemistry of carbon compounds. The old name, "organic chemistry," persists, of course, because it is useful, but without its former connotation of a life force.

Let us consider the periodic table again. We can imagine it is a fantastic sort of castle with 103 rooms in it. In each room there are collected an element, its isotopes, its ions, and all the compounds that it forms. Many of these compounds turn up in several rooms. Some are glittering and colorful; there are gases, liquids, and solids. Not every sample will endure forever, because some are radioactive. In the room for carbon we find a black, soft, greasy sheet called graphite, and a transparent crystal that breaks up light into flashing rainbows and is the hardest known substance—diamond. There are the gases CO and CO_2, a gray, very hard substance called calcium carbide, CaC_2, and a few other compounds. There is a window in this carbon room, and if we look out through it we see spreading as far as the horizon a jungle of every kind of grass, tree, flower, vegetable, creeper that has ever grown on earth, and in the branches and among the roots of this jungle, and high above it, and in the pools that water it, are all the grubs and insects, the birds and beasts, the fishes and crustaceans that have ever lived or ever will live here on earth. And all this endless proliferation, all in constant growth and reproduction—all consist of carbon compounds.

We are looking at what appears to be an unimaginably complicated confusion of matter. Can a million or so known carbon compounds be organized into families, groups, series, in such a way that the countless millions of carbon compounds we can theoretically invent fit into those categories too? The answer is an emphatic yes. In fact, one of the fascinating things about organic chemistry is that it can be organized in a way that inorganic chemistry cannot.

As the name implies, the key to the variety of carbon compounds is the carbon atom itself. There are 6 protons and 6 neutrons in the nucleus of the most common carbon isotope. It has 2 electrons in its K shell and 4 in the L shell. The four valence electrons must be imagined taking up positions in three-dimensional space around the nucleus, equidistant from one another on the surface of a sphere, as it were. The angles between lines drawn from them to the center of the nucleus are the same. Each electron is equally capable of forming a covalent bond with an electron from another atom, and thus the count of carbon's electrons can be raised to 8. The strength of these bonds is identical. The other atoms joined to carbon can be four more carbon atoms, and if each of those is linked to four more carbon atoms, and so on, that latticework creates diamond, the hardest substance. In graphite, a soft black stuff, the carbon atoms share their electrons differently. Many other elements occur in two or more crystal shapes, called allotropes of that element. If the allotropic form of carbon that is graphite is put under enormous pressure and heat, the carbon atoms rearrange themselves and form the bonds that create diamond.

As vegetation dies and decays, its complex carbon compounds break down into simpler and simpler molecules. As forest floors deepen, the underlying levels are sealed off from the air, and gradually, with increasing pressures, the process of rotting produces vast bogs of carbonaceous garbage. Upheavals in the crust of the earth bury these bogs, and far greater pressures slowly crush the whole area until it consists largely of carbon with residues of organic compounds: coal. When coal sinks deeper and is further squeezed by movements of the crust so that the pressures become phenomenally high, pure carbon in the form of graphite, and finally diamond, is produced.

Such is the picture of the natural occurrence of the element carbon. Every molecule of every living thing also contains carbon.

The compounds of carbon have been separated into three great categories: the hydrocarbons, the carbohydrates, and the proteins, with derivatives of each category forming a variety of subgroups. To think of all the materials of life and to say that there are only three basic kinds is to say something fantastic but considerably less so than to say that the whole universe consists of protons, neutrons, and electrons.

The hydrocarbons

These are all the compounds that consist of carbon and hydrogen only, in various combinations, and all the compounds which are derived from these in certain ways.

Hydrogen, as we know, forms covalent bonds easily, and the simplest carbon compound with hydrogen would be CH_4, which is called methane. The next simplest is ethane, C_2H_6. The third is propane, C_3H_8, the structural formula of which is:

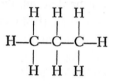

There is a whole series of increasingly longer carbon chains in which each carbon atom is linked only to hydrogen atoms until there are more than seventy carbons in the chain. Inspection reveals that they all have the same general formula, which can be written C_nH_{2n+2} where n is the number of the carbon atoms in the molecule. To calculate the hydrogen atoms in octane, which has 8 carbons: $(2 \times 8) + 2 = 18$, and thus the formula is C_8H_{18}.

Such families, in which the atoms are arranged in the same general way, are the backbone of organic chemistry and are called homologous series. Every homologous series has a common or general formula for all its members in which the ratio of the different elements is algebraically indicated.

The structural formulas for this first and simplest series are drawn in the manner that has become conventional over the years. A line in a structural formula always stands for a covalent bond—a pair of shared electrons. Structural formulas do not indicate one of the most important aspects of these molecules, that they exist in three dimensions. Not only can each atom swivel about its covalent bond, but the bonds can be bent to a variety of forces. Carbon-to-carbon bonding is remarkably flexible, yet very strong, and a great many biologically important compounds cannot possibly be represented adequately in two dimensions.

The simplest series of hydrocarbons is called the alkanes. The first five are gases at ordinary temperatures, the next fifteen or so heavier ones are liquids, and those that have more than 18 carbon atoms are solids at room temperature, white and waxy. The series is therefore also called the paraffins. They all occur in natural gases and petroleums and they can also be synthesized from other organic compounds.

If we examine butane, C_4H_{10}, the fourth in the series, we find that the carbon atoms can be arranged in a straight chain, as indicated by the structural formula:

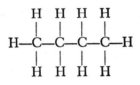

and also in a branched way:

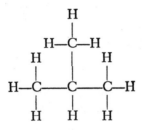

and that each has the same empirical formula. Are the two forms the same compound? Obviously they cannot be the same and in fact they are quite different both physically and chemically. Compounds with the same formula but not the same structural arrangement of the atoms are called isomers. Butane has two isomers, but the longer the molecule, the more ways there are of arranging the atoms, and the higher alkanes have dozens of known isomers. As for theoretically derived isomers that have not yet been created, decane $C_{10}H_{22}$ has seventy-five, octodecane $C_{18}H_{38}$ has over six thousand, and the theoretical number for even higher alkanes runs into the trillions.

The next simplest hydrocarbon series is called the olefins or the alkenes, with a general formula C_nH_{2n}, and the third homologous series is comprised of the alkynes, which have the general formula C_nH_{2n-2}.

In each of the alkenes two of the carbon atoms are joined by a double bond. Two pairs of electrons are shared between them, leaving only two electrons on each of these carbons for hydrogen. We have run into double bonds before, for instance in the molecule of oxygen O_2. A double bond in organic structures is represented by a double line. The alkynes have a triple bond somewhere in the chain. Any compound with a double or triple bond will be frozen, as it were, at that bond, and the carbons on either side of it will not be able to swivel. Thus the actual position of a double or triple bond in the chain adds to the number of isomers. Compounds with double or triple bonds are called unsaturated because they do not have the expected number of hydrogen atoms in them. A saturated compound has no double or triple bonds. The nomenclature of the alkenes and alkynes has been conventionalized to parallel that of the alkanes, but many of the old names remain because of their long commercial history. For instance, the new name for $CH_2 = CH_2$ is ethene, but everyone still calls it ethylene.

The three families we have defined—each of which has a general formula—are together called the aliphatic hydrocarbons. The next largest family of hydrocarbons consists of those whose carbon chains are bent into rings with the two ends joined together. They are called the alicyclic hydrocarbons, which are designated by the prefix "cyclo" added to the analogous alkane, for instance, cyclopropane. There are cyclic compounds with as many as ten carbons.

There is another group of hydrocarbons which consists of cyclic compounds based on the benzene ring. These are called the aromatic, the benzene, or the alkyl hydrocarbons, and their derivatives. They are so numerous and so important in many ways that they constitute a special study within the general body of organic chemistry.

The benzene ring, which is the basis for this complicated family that does not have a general formula, is a six-sided cyclic compound between whose carbon atoms the bond is neither single nor double but one that alternates between the two—a type called a hybrid resonant bond. The story of how this mechanism was worked out concerns some of the most important developments in chemistry during the last hundred years, and the final solution illuminated a mass of otherwise inexplicable information about molecular behavior. The empirical formula for benzene is C_6H_6, but usually the presence of the ring in a complicated molecule is indicated by a small hexagonal.

Carbon-link reactions

All these hydrocarbons, and all other types of organic compounds, whether straight-chain, branched-chain, or cyclic, react with one another and with other kinds of compounds in four general ways, at a carbon link.

First, almost any hydrogen atom on a carbon atom can be replaced by a variety of other atoms or molecules. From methane, CH_4, we can synthesize the compounds CH_3Cl, CH_2Cl_2 (chloroform), $CHCl_3$, CCl_4 (carbon tetrachloride, the cleaning fluid). This process is called replacement.

Secondly, the carbon compounds can be made to open their double and triple bonds between carbon atoms, and hydrogen, or other atoms or molecules, can be added to those carbon atoms. Hydrogenation is the industrial name for the most common of these processes, and is best known when used with unsaturated fats and oils, which are higher alkene derivatives. Hydrogenation turns a liquid oil into a solid fat. These addition reactions obviously cannot take place with saturated compounds.

Third, any unsaturated organic compounds can, by breaking the double bonds between carbon atoms in this manner, add themselves together into enormously long chains, a process called polymerization. Saturated compounds can, through replacement, also polymerize. Polymerization is the basis of the manufacture of all plastics and of many pharmaceutical complexes.

The fourth general type of reaction that all organic compounds can be made to undergo is a simple breaking-up process at the carbon links. Out of large molecules, chain or cyclic, saturated or unsaturated, smaller ones can be made, and if the raw material is petroleum, for instance, the process is called cracking. Thus, all these reactions occur at the carbon atom in organic compounds.

Free radicals

Every organic molecule, no matter how complex, can be made to lose a single hydrogen atom from one of its carbon atoms and so become free to link itself to any other compound or element that needs a single covalent bond. An organic molecule with such a free bond cannot exist by itself. If two collide, they link at once. These fragments of organic molecules, lacking only one hydrogen atom, are called free radicals, to distinguish them from the ionic radicals of inorganic chemistry.

Methane's free radical is CH_3—, or methyl. This group of atoms, the methyl free radical, which cannot ever be isolated by itself because two of them at once link into CH_3—CH_3, or ethane, turns up in a vast number of compounds and its name is part of their name, as in methyl alcohol. CH_3CH_2—is the ethyl radical, and it too occurs in a very large number of compounds as a branch chain. The alkyl or benzyl radical is C_6H_5— and thus methyl benzene is CH_3—C_6H_5.

Every organic molecule, without exception, can be thought of as a potential free radical. Thus an unlimited number of compounds can be imagined by putting together in different arrangements a relatively small number of free radicals. Much of the research in organic chemistry has been the attempt to synthesize theoretically possible molecules out of existing free radicals.

In writing general formulas, which are very common in organic chemistry because of homologous series, any free radical can be represented by the letter R. Thus R—CH_3 can be CH_3—CH_3, or C_2H_5—CH_3, which is methyl ethane or ethyl methane, and also propane, or C₆H₅—CH₃, which is methyl benzene or benzyl methane. R can be any free radical of countless other compounds.

Functional groups

There is another category of groups of atoms which function as a unit in chemical reactions and which also cannot exist independently, and these are called functional groups. There are less than a dozen. They are attachable to carbon atoms in organic compounds of all kinds, and they have their own special properties which they add to whatever molecule they bond into. An analogy is the special hydrogen in acids, which the acid molecule can give up to a base.

One of the most important of these functional groups—always bonded to a carbon atom—has the formula —OH, but this is not at all the familiar hydroxide OH⁻ ion we discussed in Chapter 18, which accepts protons. This —OH is the hydroxyl group, it does not have any electrical charge on it, nor can it be found apart from a covalent linkage to a carbon atom in an organic molecule. Any hydrocarbon that has a hydroxyl group attached to it is called an alcohol. A molecule can have several of these hydroxyl groups attached to several carbon atoms, and it is then called a polyalcohol, some of which are called saccharides. If the free radical is R, then R—OH is the general formula for all primary alcohols. Ethyl alcohol is CH₃CH₂OH, the intoxicating spirit in all fermented liquids. Methyl alcohol is CH₃OH, wood alcohol, a poison.

Another functional group just as important as the hydroxyl is the carboxyl, which is the acid part of any organic acid:

Vinegar is acetic acid, CH₃COOH.

Other functional groups attached to radicals produce groups of compounds known as aldehydes, or ketones, or esters, or ethers, or amines, or amides. For example, an aldehyde is an organic molecule, simple or complex, that has attached to one or more of its carbon atoms this group:

If the molecule is large enough, a mixture of functional groups can be hung on the carbon spine and each one will always exhibit its properties in chemical reactions that involve that particular functional group. The naming of these complicated compounds is a headache, and a continuous revision is going on among organic chemists concerning the grouping of such compounds under identifying names. For example, many compounds are esters, ketones, alcohols, and acids, all at the same time.

Carbohydrates

We can now proceed to the second great subdivision of organic chemistry, which is called the carbohydrates, and in which every molecule contains, in addition to carbon, hydrogen and oxygen atoms in the same proportion as in water, two hydrogen atoms for every oxygen atom. One of the simplest and most important is glucose, which has an aldehyde group at one end and five alcohol or hydroxyl groups along the chain, so that it is called a polyalcohol and also a monosaccharide. The formula is $C_6H_{12}O_6$, and this is exactly the same as the formula for fructose, which has a ketone group instead of aldehyde. If fructose and glucose, two isomers, are joined with the loss of a molecule of water, the resulting disaccharide is called sucrose, $C_{12}H_{22}O_{11}$, the white granular substance we buy in the grocery store as sugar.

Glucose molecules can be joined end to end in a polymerization process until about 3000 such units are strung together. They will then constitute a single molecule of a polysaccharide called starch, the same starch we find in cereals and potatoes and rice and the starch called glycogen, which is found in flesh. The formula for it is $(C_6H_{10}O_5)_xH_2O$, where x is 3000.

If glucose molecules are joined until there are almost six thousand of them locked together into a single long strand with complicated side branches, we have cellulose, another polysaccharide, which is the structural substance of vegetable matter.

Plants and animals produce starch and glycogen, and when we eat these compounds enzymes, which are organic catalysts, break them down in the stomach to sugar units, which are then used by the body in various ways. Many insects do the same thing with cellulose. Enzymes in the stomach of wood-boring termites break down and convert the wood to sugar. Enzymes break down fruit sugars into alcohol. If we further oxidize alcohol, or a sugar, either by burning it in the air or by metabolism in our bodies, the end products are water and carbon dioxide, and heat energy.

The same carbon-dioxide gas and water vapor are recombined by living green plants and by tiny organisms in the ocean, into sugar molecules, and then these are polymerized into starch. The reaction here is endothermic and requires enormous amounts of energy to keep it going, and a catalyst is necessary to activate the reaction. The energy is the sun's light and heat, the catalyst is chlorophyll, and the process is called *photosynthesis*. All the heat given off when wood is burned, or when we eat bread, comes from the energy that carbon dioxide and water took in from the sun's radiant energy when they combined.

Fats

The next large group of organic compounds are the fats, whether solid or liquid at ordinary temperatures. These are esters of glycerol and of fatty acids, and there is no space in an introduction to elaborate this briefly yet meaningfully without discussion of esters in general and their formation from carboxylic acids. But it is useful to remember that the unsaturated esters in liquid fats, such as olive oil, peanut oil, corn oil, can be hydrogenated to the saturated form and that then they become solid at room temperatures. But the most important industry that uses fats, whether saturated or not, boils them with an alkali to produce glycerine and the metal salt of the fatty acid, which is ordinary soap.

Proteins

The last large family of organic compounds is the proteins, composed of carbon, hydrogen, oxygen, and nitrogen, and including such compounds as enzymes, vitamins, and hormones. There are literally as many kinds of protein as there are kinds of tissue in all living organisms. They are extremely complicated, enormous, and enormously convoluted, consisting of twisted and cross-linked chains, and it is their interlocking shape that gives cells their strength. These molecules actually twist into each other's jagged, indented spiraling shapes and become tangled and coiled, and thus not only create the strength of hair, skin, bone, muscles, but also their flexibility, resilience, and stretchability. Furthermore, the key to every cell, the active ingredient within the cell that is the organizer of the cell's existence and of its function in the living body, is also a protein.

All proteins are polymerized from smaller molecules called amino acids. Only twenty-odd amino acids are known, but these can be put together in endless variety, much as the twenty-six letters of the alphabet can be put together to form countless combinations that spell the different words of all the Western languages and any others we choose to invent.

The vitamins are grouped together more because of the similarity of their function in the body than because of any consistent similarity in their structure.

Drugs, synthetic fibers, and research, pure and applied

Another grouping of organic compounds is determined by the use to which they are put by man: the drugs. Pharmaceuticals are a separate branch of chemistry because they are employed against disease and not because they are similar in any way, though many of them can be grouped chemically. Yet often very similar compounds react so differently with the body's compounds that one may be a poison and the other a necessary substance for life. The use of drugs seems essential to all men, and the search for miracle pills is older than the alchemist's dream, older than Egypt's priestly magic, older than a witch doctor's rattling bones and smoking fire. A man with sinus trouble will try anything that is offered him, even if he knows there is no cure. Herbs, poultices, vaccines, syrups, and ointments become fashionable, then vanish from the druggist's shelves, then return again, throughout history. The multi-billion-dollar pharmaceutical industry is searching for new drugs constantly.

The modern chemical industry which actually squeezes drugs out of first place in terms of yearly profits is the synthetic-fiber and plastic field. Research revealed the patterns into which sugar molecules are polymerized by living organisms, and then techniques were developed for synthesizing sugar molecules into substances very much like cellulose. In the same way compounds were synthesized which are very much like natural rubber and various natural resins.

The word "synthetic" has acquired the connotation of spurious, false, untrue, and this is not correct. Nylon, a synthetic fiber, is true unto itself although its molecular formula closely resembles that of silk and it can be called an imitation silk. If it were possible to manufacture real silk, it would be done only if the process made silk more cheaply than silkworms. Still, there is the research challenge: why can't we make genuine silk? The molecules are too huge and complicated to be put together without small deviations from the original formula, but sometimes these deviations produce stuff that is more useful in our industrial age than the natural fiber. Nylon's strength cannot be matched, nor can its imperviousness to rot be found in nature.

When it was discovered that natural fibers which had clothed and housed man since the beginning of time could be replaced by polymers manufactured from wood pulp and petroleum, the economics of sheep raising and cotton growing were immediately affected, but a much more profound revolution had already begun in our culture, and it concerned the rising cost of modern research. In every branch of science the universities were receiving increasing sums of money from the government, from industry, and from the foundations, to do work which a generation ago the universities had done as a matter of course on their own budgets, and industry was beginning to invest heavily in its own research plants. Today the research picture reveals a truly fantastic change from the 1930s, and the differences between basic research, practical research, applied science, technology, engineering, have become a matter of point of view. One point of view cannot see any difference between pure and applied science that is not just a conventional mode of speech, but even the point of view which does distinguish between a search for laws with which to create order in the ocean of information-data and a search for the solution to some specific problem admits that the distinction is fast fading. The springs of imagination that used to feed theoretical work on abstract principles today feed work on the practical applications of scientific knowledge too. It is impossible to decide how much pure research there should be and how much applied. Science has become an activity that is not merely profitable to some industries, but is vital for the survival of mass populations.

On this portentous sociological note, let us refocus our imagination to bring back into view a final glimpse of the molecular nature of matter.

22

The Smallest Clot of Molecules

Colloidal dispersions defined

THE molecules of a solute are in constant random movement, dispersed evenly among the constantly moving molecules of a solvent. Obviously a study of the forces between them will involve a study of actual contact between what, for want of a better word, are known as molecular surfaces. For various reasons, since World War II, a whole new field has developed which is concerned with molecular surfaces, and solutions are only a small part of it. The wetting of surfaces, adhesion and absorption, the processes of sharpening and polishing ultrafine points, must all be explained in terms of electromagnetic forces between different molecules, forces that do not bring about chemical change and yet involve the electron clouds of nuclei. Metal alloys which with a slight change in percentage composition, or just in the temperature of annealing and rate of cooling, acquire startlingly different properties, all kinds of monomolecular layers which act not just as binders or repellents but also as screens and sieves and mirrors, surfaces which ionize under special conditions or become conductors or nonconductors, are a few more examples of new targets in solid-state physics. The field has expanded so rapidly that in the last ten years specialized areas have developed within it, mostly relating to electronics, that catch-all word which includes any movement of electrons for the purpose of information under controlled and nonchemical conditions.

Between these unimolecular properties of matter and its properties in bulk, there is a region of molecular activity which belongs to neither scale, called the colloidal state of matter, thus suggesting that it is as basic an association of particles as gas, liquid, and solid are.

Imagine that you pour a water solution of a compound into the water solution of another compound and that the two compounds react to produce a solid that is insoluble in water. As this product is formed, its molecules collide with one another and, being insoluble in water and thus repelled by water molecules, they cling together in colonies until throughout the solution tiny kernels of the insoluble solid begin to appear. A cloud of crystals growing and coarsening and settling out is one of the most commonly desired results of chemical reaction because the solid can be filtered out.

When these seedlings are still submicroscopic they have some of the kinetic energy of the molecules, but as they grow they move less and less, and as the weight of each kernel increases a point is reached where the seedlings will be motionless, except for battering by high-speed water molecules which roll the particles a little this way and that. Eventually the crystals will grow to such a size and mass that gravity will slowly pull them down. By then they will be visible to the naked eye.

But imagine a glass of water containing a mist of these incipient crystals, each consisting of a score, or a hundred or so, molecules, and too small to be seen in the most powerful optical microscope, and imagine that for some reason they have stopped growing. This invisible cloud of clots is called a colloidal dispersion.

The properties of colloids have fascinated scientists since ancient times, though their structure could not be deduced. The tanning of hides, the dyeing of cloth, paper-making, glass-making, the grinding of paints, the hardening of clays, are only a few operations which are concerned with colloidal dispersion. The word was first used in the nineteenth century to designate a substance which, instead of solidifying from solution into crystals, turned into amorphous glues, jellies, slimes. Like so many definitions, the term was expanded to include more and more and different phenomena, until today it has nothing to do with glue or with special compounds of any kind.

Colloidal dispersions can be produced out of any mixture of compounds or elements; it is simply a matter of creating, and then of arresting, the growth of submicroscopic particles of one substance in another. Submicroscopic gas bubbles in a liquid produce foam; tiny droplets of liquid in a gas produce mists, or aerosols; droplets of liquid in another liquid, such as fat in milk, are called emulsions; invisible solid particles in air are called soot or smoke, and those in a liquid are called sols; gels are semiliquid coagulations of sols and are like jellies. All of these are colloidal dispersions.

The substance in which the particles are suspended is the dispersion medium, while the particles themselves are called the disperse phase, and the two together are a colloidal system. The word "colloid" is used loosely to indicate the system, and sometimes to indicate only the dispersed phase. The terminology is rather recent and some of the old names are still useful—such as "jelly," which is not defined in the new vocabulary. Some old words have been reshaped, as "sol," for a solid phase dispersed in a liquid phase.

Since in a colloidal system the particles are clots of molecules, they cannot be called solute particles and the whole cannot be called a solution. Yet, like solutes, colloidal particles are submicroscopic, still have a certain movement, and do not settle out. In a heterogeneous mixture the particles always sink, owing to gravity—for instance, dust in air, or the nonfat part of milk—but colloidal dispersion persists in maintaining a uniform distribution of its invisible clots. Another property of true solutions is that the solute cannot be filtered out of the solvent, both being molecular in size and both passing through any kind of filter. A colloidal system's particles are small enough to pass through the finest filters known.

Then how can we distinguish a colloid from a solution? One test is with a beam of light. A true solution will let light pass through without reflecting it. The particles of a colloid, however, scatter light rays, and a beam of light directed at a glass containing a sol will light up its path through the sol. This is called the Tyndall effect and is proof that something larger than molecules is present. Brownian movement of submicroscopic particles can be studied in an apparatus which discovers incessant flashing of reflected light, as though by the twirling and bouncing of invisible mirrors. All colloids produce the Tyndall effect and Brownian movement, and so these can be considered properties that belong to the colloidal state exclusively, together with the property of remaining in suspension and of not being filtrable.

A little reflection will reveal that the essential question here is: if molecules have begun to form clots or droplets, why does the growth process stop at colloidal size? There are two answers, and each is both simple and complicated, and again each is electromagnetic in nature.

The first explanation is that each seedling crystal or droplet in a system acquires the same kind of electrical charge as all the others, either positive or negative, so that in any one system all the particles will have only one kind, and the seedlings therefore repel one another, and will be unable to continue coagulating. The charges will push them apart as they approach one another.

This electrical charge can come from ions present in the solution and sticking to the clots which are no longer darting about, or they can develop in the structure of the clotting molecules themselves. If the particles have acquired a negative charge and positive ions are added to the system, they will neutralize the colloidal charges and the particles will resume their growth and precipitate.

The second inhibitor to the growth of particles is a strong attraction between them and the molecules of the medium, which attach themselves to the seedlings and act as physical buffers to prevent further growth. The addition of a substance that will tempt the medium molecules away from the seedlings will permit them to resume their growth through collisions with each other. Heating and centrifuging will often force coagulation.

A colloidal dispersion carrying an electrical charge will not conduct electricity, because all the particles have the same charge but, when a current is passed through the system, the particles will gradually move toward the oppositely charged electrode and become neutralized. If there is a mixture of colloids, as is often the case, especially with aerosols, those positively charged can be separated from the negatively charged ones by applying an electrical current; both will be neutralized at the electrodes and will grow into coarse size. The phenomenon is called electrophoresis.

Creating and destroying colloids

In industry and pharmacology a colloidal condition is often desirable, and two general methods can be used to prepare it. The first is to grind, mash, pulverize bulk matter into submicroscopic size, something that is not possible with all substances. Rotating drums filled with rolling and tumbling steel balls is a common piece of equipment for this operation. The other method is to begin with a pure solution, allow crystallization to start, and then arrest it at the right point with the addition of buffering compounds or ions.

Many aspects of the life process are organized around the colloidal state. Digestion, or peptization, is the process of reducing chewed-up food to colloidal size, while cell formation is the process of congregating and sticking together molecules into colloidal-sized clots.

The study of cell structure, which has become increasingly fascinating with the development of the electron microscope that actually casts shadows of giant molecules upon photographic film, is a study of matter at the colloidal scale of molecular activity. One does not have just a heap of molecules at one instant and fully organized tissue at the next instant. Obviously there must be a congregating, a selecting and melding, a steady growth of tissue stuff from basic molecules. Thus, all the vast variety of tissue structures are today seen partly as colloidal systems which can swell, shrivel, stretch, whose walls are permeable to some molecular and ionic compounds and not to others. Therefore chemical reactions inside and outside the walls of cells cannot be understood unless the walls are seen at colloidal levels.

At this scale of magnitude, surface contact between molecules is fascinating. Tiny particles, whether in the liquid or solid phase, have surface tensions which are quite different in scale from those of bulk matter. To illustrate this, imagine a cubic inch of chalk, i.e., a cube with each side 1 inch long. The total surface area of the cube is 6 square inches, there being 6 sides to the cube. Imagine the cube cut into cubes that are $\frac{1}{10}$ of an inch along each side. Each of these smaller cubes will have a surface area of $\frac{6}{100}$ of a square inch, but there will be 1000 such cubes, so that their total surface area will be 60 square inches. If each of these little cubes is now cut up into cubes $\frac{1}{100}$ of an inch along a side, the total area will be 600 square inches. As a piece of matter is ground down, its total surface area increases with startling rapidity and becomes enormous by the time a colloidal state is reached. The tensions of surfaces, therefore, become increasingly important, and the study of surfaces has become an intensely busy part of research.

The two most prevalent colloidal systems are sols, which are solids in liquids, and emulsions, which are liquid droplets in liquids. The stabilizing of an emulsion illustrates how important the knowledge of surface physics is in this field. If oil is shaken up with water the oil will be broken up into fine droplets but, as soon as the shaking stops, the droplets begin to rejoin and soon the mixture of oil and water will separate into two layers. But the mixture can be stabilized by emulsifying agents, for example, soap. One end of a soap molecule is acidic and the other end is inert. The acid end is highly soluble in water, the inert end in fats and oils. The molecules of soap range around a droplet of oil like pins in a pin cushion, and sink their inert heads into the droplet, leaving their acidic tails in the water. The droplets are now effectively shielded from all the other droplets by the bristling acidic tail ends. Such is the mechanism with which soap and detergents remove dirt. They literally lift it away from the surface to be cleaned. Cleanliness is an emulsifying act.

Photographic film is also an example of successful emulsification and stabilization on an industrial scale. The number of stabilized colloids in pharmaceuticals and food products is legion.

Aerosols, or solid and liquid particles in gas, are important mostly because they are the undesirable products of all kinds of furnaces and engines. Various methods of filtering industrial smoke have been developed, none wholly successful; methods of washing exhaust fumes are generally too expensive; improving burners and smokestacks to provide total combustion has not been the answer. A device called the Cottrell precipitator seems to be the cheapest and most reliable way of removing colloidal dispersions from gases.

This equipment, which is an application of electrophoresis, consists of a series of baffles through which the smoke must coil, parts of the baffles being electrically charged. The positively charged colloidal particles are trapped in one place and the negatively charged ones in another place. As soon as the charges are neutralized by the electric plates, the particles snowball into coarse pieces that can be shaken together and collected and handled in the manner of ordinary rubbish. However, this method does not remove colloidal dispersions which are not electrically charged. Most localities that have laws against pollution prescribe Cottrell precipitators for industrial furnaces but, unfortunately, it remains considerably cheaper to let the oily mist and soot float off on the air than to collect it.

In some industries valuable products are carried off in aerosols, and then every effort is made to salvage them. For instance, the loss of precious metals during the smelting of various ores can be very serious.

An important point in the study of colloidal dispersions is that impacts between independent molecules are not the same as collisions between clusters of molecules. Different forces are brought into play and special properties emerge, as the size of clusters increases and the curvature of their surface becomes larger and the total amount of exposed surface in the system diminishes. We are constantly dealing with averages.

Many fantastic aspects of colloids cannot be examined here—

such as dialysis, which is the purification of a system, whether in living organisms or not, through semipermeable membranes. In a tree, whose woody strength also must grow through a colloidal stage, and which keeps itself flexible as the wind bends it, tons of water are lifted through the capillaries of the trunk and branches to the twig ends and leaves. The power to raise this water from the ground is provided by surface tensions between molecules, which also regulate the tree's flow of juices according to the seasons and the weather.

Library shelves are bending from more and more publications on colloids and practically all of them are comprehensible only to researchers. Behind every bit of casual reference in this chapter there lie many man-years of patient observation, laboratory experimentation, and mathematical operations. The general point of view taken today is that the vast spectrum of invisible activity in the world, and probably even in the universe, is colloidal in nature, and that the human body is an unimaginably complicated balance between colloidal systems.

23

Science Belongs to Everyone

Reflections on things past

As our ordered information on matter deepens and as our use of that knowledge also deepens in every field of study, our concepts of matter inevitably change. A ceaseless process goes on that is not difficult to grasp. In a broad sense a scientific law becomes a useful tool for performing specific operations and calculating predictions. As measurements in an area are refined, the laws often have to be redefined, sometimes quite altered in order to embrace the new discoveries. The illumination for this may come from a source that until then was not related to the field where the new law is needed.

Not only laws but theories also are incessantly being reshaped to account for new and often contradictory data. For example, the theory that atoms were ultimate particles of matter worked with such magnificent success in every branch of science that when the electron was discovered it was almost impossible to believe that it really had come out of the atom. Radioactive elements which hurled subatomic particles out of their atoms were an even more difficult piece of reality to contemplate with Dalton's atomic theory, yet the concept that atoms were the smallest recognizable particles of the elements had become a certainty. The whole constitution of the atom had to be revised radically with new theories based on new observations in many different kinds of research.

What shall we think then of the faith scientists had a century ago that the atom was indivisible? Whatever we conclude, we must think the same of ourselves today. We too have worked out wonderful ideas that are prying open one mysterious piece of reality after another, and some of these ideas are bound to appear primitive, even wrong-headed, in a very short time. The speed of light, we now say, is always the same in empty space, and this sameness allows us to explore the photographs our telescope cameras take of the starry heavens with a calm conviction about certain things that light can and cannot do. The speed-of-light-is-always-the-same is a piece of theory-law without which it would be impossible to conduct modern research. But it is based exclusively on the fact that we have never caught light behaving otherwise. If it turns out that we can catch light at other speeds in empty space, it will not invalidate one single item of information that we have collected with the old faith in light. Furthermore, the always-the-same theory would still be true within the limits in which we have been using it, and it would always remain useful as an approximation. But obviously, our concepts of light would have to change.

This is the great difference between science and philosophy or religion. The latter are constructions of thought designed to guide man according to certain assumptions concerning his worth in eternity, or his universality, and they make permanent and un-equivocal statements about man's earthly lot and thus about human values, ethics, the good society. The unchanging pattern of any philosophy's logic concerning the nature of existence, the stiff dogma of any religion's view of existence, is at startling variance with what we find in science. There is a constant meddling with the laws, the theories, the truths of previous generations, rather than a striving to bring about ever deeper confirmation of those truths.

It is the very essence of scientific work that there is no central authority, no organization, no individual, living or dead, whose function can be compared to that of prophets, founders, church, hierarchy, government, oligarchy, school, priesthood, elected or appointed officialdom of any kind. There is no body of knowledge,

no holy writ, no practice, no fountainhead, no object of venera-
tion that is cherished for its intrinsic power to formulate truth
above the discoveries of the least-known practitioner. All are
equal in science at a level unknown in any other human coopera-
tion. The reader already knows more about the nature of matter
than did the greatest scientists who died fifty years ago. There is
a vital continuity of growth, an evolution, in the history of science
which has no counterpart in any other social activity.

The various schools of logic not only stand apart from one another but are often dramatically locked in conflict that will never
end. It is even more profoundly true of the various religions that
they are all mutually exclusive in their revealed truths. One
chooses to join one congregation, not two or three, and one's
choice is difficult to analyze. It is a matter of faith, as all religions
insist that it is. One cannot be both Catholic and Jewish. Similarly one cannot be both Aristotelian and Platonic, or both exis-
tentialist and Epicurean. In science there are, true, occasionally
so-called schools of thought concerning a phenomenon, but all
concerned are quite certain that sooner or later one of the schools
will be vindicated by experimental proof, and the losing school
will then vanish from the scene, not remain as a rival. In philosophy, if a single axiom is changed, a new school is set up and, though
it may weaken the popularity of the old school, it can never invalidate it. Each has its own logical arguments.

The splintering of religions and the fierce loyalty of each splinter
group to its measure of difference from all others is the sorrow
of all splinter groups. But one cannot have an evaluating opinion
concerning the latest figure for the speed of light in a vacuum
or the most recent measure of an atom's diameter. The law of
mass action is not a finer law than the law of gravity. Heresy is
impossible in science, though a lot of scientific work is useless,
a lot of scientific discussion is vapid, and a lot of stupidity and
wrong-headedness flourish. All is cleansed or eliminated as a matter
of course by the endless experimentation that goes on all over
the world. The greatest network of international communication
that has ever existed is one of the most important aspects of scien-

tific progress, but it is not essential to science, which, without such collaboration, would still advance knowledge of nature wherever men found pleasure and not necessarily profit in the empirical point of view.

Since science-knowledge is based on the actual measurement of observable events, it is the only method of thinking that so far has outstripped all other methods in predicting the future—not all futures, but those within its realm. A modern factory is inconceivable unless we understand that it is built on the faith that machines will perform precisely their appointed tasks. Watt's steam engine, like the simplest contrivance of antiquity, was mysteriously wonderful because it was predictable in its movements. Fountains and water wheels, the loom, even the sundial, still fascinate us. Such predictability is somehow refreshing to us, enslaved by the miserable cat-and-mouse game going on incessantly between our dreams and our sense of reality, between our ambitions and our fate, between our faith in continuity and our recognition of accident.

Yet paradoxically our faith in science rests not in the laws themselves, but in the apparatus of thought which is free to improve the laws. It is never a theory that demolishes another theory—it is a fresh batch of ordered information collected by scientific tools. Furthermore, though most old theories are melted into the new ones, those that have become utterly useless are remembered as being perfectly valid for their time, related to the information that their age possessed.

And so we have reached the end of our introduction to chemistry. We began with the enormous variety of our sense impressions of objects, their properties and behavior. We analyzed the universe into scores of different categories, beginning with air, earth, water, and fire, and ending with protons, neutrons, and electrons, and we investigated the nature of chemical change.

The field of nuclear reactions—fission and fusion—has been omitted entirely, since this momentous subject has been explained so often in print at every level of simplification. But the reader will realize that the so-called atom bomb, atomic reactors, atomic power, atomic fission and fusion are all misnomers, as all are concerned with nuclear, not atomic, reactions. The word "atom" connotes the smallest particle of an element, consisting of a nucleus surrounded by shells of electrons, which determine the chemical properties of the whole atom. In nuclear physics all the electrons —not just those in the valence shells, but in all the shells—are simply brushed aside as though they were irrelevant to the behavior of the naked nucleus. Thus, nuclear reactions have nothing to do with chemical reactions. But it is fitting to mention here that nuclear research today points to the possibility that the atoms of all the elements are being manufactured in the stars, out of protons and electrons and neutrons and the other slivers of matter-energy. We have discovered how the sun manufactures helium from hydrogen, and once this was understood, the rest could be worked out theoretically.

The future

All the knowledge of the nature of matter is now also being focused on the living cell and on the instant when a heap of molecules mysteriously begins to organize itself into protoplasm. At the same time the mystery of lifeless space pulls our imagination and our efforts in another direction. Test-tube life and a landing on the moon seem equally near. Whether or not these ancient desires that have no moral significance in the traditional sense will be realized for moral purposes or for sheer joy in the doing does not seem to matter at the moment. Something is guiding our energies that will not be denied. But we must still eat together, still make love, breed, suffer and weep, laugh together, willy-nilly, and die in each others' arms. No science can eliminate society, only man can, and the future still belongs to those, it seems, who accept the profoundly troubling fact that being human means to believe that we are free to choose our human destiny.

Index